1994

SECOND EDITION

Description and Sampling of Contaminated Soils

A FIELD GUIDE

J. RUSSELL BOULDING
Boulding Soil-Water Consulting
Bloomington, Indiana

LEWIS PUBLISHERS
Boca Raton   Ann Arbor   London   Tokyo

Library of Congress Cataloging-in-Publication Data

Boulding, Russell.
    Description and sampling of contaminated soils: a field guide / by J. Russell Boulding. — 2nd ed.
        p.   cm.
    Includes bibliographical references (p.   ).
    ISBN 1-56670-050-7
    1. Soil pollution.   2. Soils—Sampling.   I. Title.
TD878.B68   1994
628.5′5—dc20

                                                                                        93-6127
                                                                                        CIP

© 1994 by CRC Press, Inc.
Lewis Publishers is an imprint of CRC Press

No claim to original U.S. Government works
International Standard Book Number 0-56670-050-7
Library of Congress Card Number 93-47072
Printed in the United States of America   1 2 3 4 5 6 7 8 9 0
Printed on acid-free paper

CONTENTS

* Indicates that guidance for interpretation of observations and measurements can be found in Cameron (1991).

FORMS

TABLES

FIGURES

ACKNOWLEDGEMENTS

This is a revised and expanded edition of **Description and Sampling of Contaminated Soils: A Field Pocket Guide** (EPA/625/12-91/002) prepared for EPA's Center for Environmental Research Information, Cincinnati, Ohio, and the Exposure Assessment Research Division of EPA's Environmental Monitoring Systems Laboratory, Las Vegas, Nevada under a contract with Eastern Research Group, Inc. (ERG), Lexington, Massachusetts.

I would like to thank the individuals who made the first edition possible: Carol Grove and Jeffrey van Ee (EPA) and Heidi Schultz (ERG). I would also like to express my appreciation for the helpful review and comments by those who reviewed earlier version of the first edition: Noel Anderson (Geraghty and Miller, Madison, WI), Robert Breckenridge (Idaho National Engineering Laboratory, Idaho Falls, ID), Roy Cameron (Lockheed Engineering & Sciences Company, Las Vegas, NV), H. Raymond Sinclair, Jr. (Soil Survey and Interpretations and Geography Staff, USDA SCS National Soil Survey Center, Lincoln NE), and Bobby Ward (State Soil Scientist, USDA Soil Conservation Service, Indianapolis, IN).

Bobby Ward and John Witty (National Leader, Soil Classification Staff, USDA SCS National Soil Survey Center, Lincoln, NE) assisted greatly in revisions to the second edition by making available the most recent versions of SCS guidance documents on soil description and classification. Thanks, also, to Lou Langan for reviewing the expanded criteria for hydrologic soil groups in Table 2-3.

J. Russell Boulding first began working in the environmental field in 1973 when he helped set up the Environmental Defense Fund's Denver Office, and has been a free-lance environmental consultant since 1977 when he established Boulding Soil-Water Consulting in Bloomington, Indiana. He has a B.A. in Geology (1970) from Antioch College, Yellow Springs, Ohio, and an M.S. in Water Resources Management (1975) from the University of Wisconsin/Madison. From 1975 to 1977 he was a soil scientist with the Indiana Department of Natural Resources and mapped soils in southern Indiana on a cooperative program with the U.S. Soil Conservation Service. Since 1984 he has been Senior Environmental Scientist with Eastern Research Group, Inc. in Lexington, Massachusetts.

Mr. Boulding is the author of more than 100 books, chapters, articles and consultant reports in the areas of soil and ground-water contamination assessment, geochemical fate assessment of hazardous wastes, mined land reclamation, and natural resource management and regulatory policy. From 1978 to 1980 he served as a member of the Environmental Subcommittee of the Committee on Surface Mining and Reclamation (COSMAR) of the National Academy of Sciences and as a consultant to the NAS Committee on Soil as a Resource in Relation to Surface Mining for Coal. Mr. Boulding is an ARCPACS Certified Professional Soil Classifier and his professional memberships include the Soil Science Society of America, International Society of Soil Science, Association of Ground Water Scientists and Engineers, and the International Association of Hydrogeologists. Since 1992 he has been a member of the American Society for Testing and Materials' Committee D18 (Soil and Rock) and active in subcommittees D18.01 (Surface and Subsurface Characterization), D18.07 (Identification and Classification of Soils) and D18.21 (Ground Water and Vadose Zone Investigations). In 1993 he became chair of D18.01's Section on Site Characterization for Environmental Purposes.

FOREWORD TO SECOND EDITION

This second edition differs from the first edition in two major respects:

1. The larger format has allowed inclusion of detailed checklists and tables in Appendix C to assist in identification of methods for field, laboratory, and estimation of site and soil parameters. An abbreviated version of this material appeared in Cameron (1991).

2. It incorporates significant changes and improvements in the way soils are described and interpreted since the first edition. The most important of these changes relate to description of redoximorphic soil features and a more precise approach for defining aquic conditions in soil (see Appendix D). Although they have not been officially adopted by SCS, this edition uses criteria for placement of soils in hydrologic groups (Section 2.4), and definitions of soil drainage classes (Table 3-10) proposed by Langan and Lammers (1991).

At the time the first edition was written, it was expected that the new edition of the **Soil Survey Manual** would be published in 1991, and was cited with a 1991 date. In early 1993 the page proofs were being checked at SCS's National Soil Survey Center in Lincoln, Nebraska, so the manual in cited with a 1993 date, even though it had not actually been published at the time revisions to the second edition of this manual were completed. SCS kindly provided disk files of final manuscript so that all tables in this guide that come from the **Soil Survey Manual** could be double-checked for accuracy.

The large number of SCS documents that relate to description, mapping and classification of soils can be confusing to individuals outside the agency. A new discussion at the beginning of the reference section should explains the purpose and relationship of different SCS handbooks and publications.

HOW TO USE THIS GUIDE

This guide describes many field methods and procedures that can be used for (1) preliminary site reconnaissance, (2) detailed site and contaminant characterization/sampling for transport/fate modeling and risk assessment, and (3) for remediation selection and design.

All methods and procedures described in this guide are simple and inexpensive. When used early in site reconnaissance, site characterization, or remediation projects, the methods in the guide may reduce project costs by providing a basis for more efficient application of more complex and expensive field methods, when they are needed.

This guide has also been designed to serve as a companion to EPA's **Guide to Site and Soil Description for Hazardous Waste Site Characterization** (Cameron, 1991), which also serves as the basis for the site and soil components for metals of the Environmental Sampling Expert System (ESES) under development by EPA's Environmental Monitoring Systems Laboratory, Las Vegas (see Section 1.6).

Use for Preliminary Site Reconnaissance. If a soil survey prepared by the Soil Conservation Service (SCS) of the U.S. Department of Agriculture is available for the site, this guide in combination with Cameron (1991) can be used to develop preliminary interpretations concerning the potential for site and soil conditions that facilitate or inhibit contaminant transport.

Use for Detailed Site and Contaminant Characterization:

1.   To assist field personnel in preparing for a visit to a contaminated site by providing checklists to ensure that no documents or equipment are accidentally left behind (see Forms 1-2 and 1-3 in Chapter 1).

2.   To provide a concise, but comprehensive reference source for methods of describing and analyzing site and soil characteristics in the field that require only visual/tactile observation or very simple equipment. Chapter 2 provides this information for site characteristics. Chapter 3 provides this information for soil characteristics, placing special emphasis on soil description procedures of the U.S. Soil Conservation Service. Abbreviations and codes that can be

used for specific soil features are suggested to facilitate note-taking. Where soil conditions favor use of a soil probe (no coarse fragments), description procedures outlined in Appendix A.1 may be useful for characterizing soils at a site prior to sampling for detailed chemical characterization.

3. To assist in selecting and obtaining alternative soil sampling equipment if unforeseen problems at the site prevent use of sampling procedures specified in the Soil Sampling and Quality Assurance Plans. Chapter 4 provides information on sampling equipment characteristics. A series of Appendices describe some standard soil sampling and handling protocols that may help address quality control concerns related to alternative procedures that may be required by unforeseen site conditions.

4. To facilitate use of EPA's Environmental Sampling Expert System (see Section 1.6).

Use for Modeling and Remediation Selection/Design. When site soil parameters that are required for modeling, or remediation selection/design, are known, the appropriate sections of this guide can be used for data collection. Appendix C provides checklists and tables for identifying soil parameters of interest.

Note for User's of the First Edition

The pocket-size format of the first edition of this guide is well-suited to carrying in the field, and those with first editions may wish to continue to do so[1].

---

[1] The first edition is still in print, and available at no cost from the U.S. EPA Center for Environmental Research Information (ORD Publications, P.O. Box 19963, Cincinnati, OH 45219-0963; 513/569-7562).

The following corrections and changes have been made in the first edition (including an indication of the type of change):

Table 2-2 (minor revision)

New Table 2-3 with expanded criteria for soil hydrologic groups

Table 3-1 (notes added to k and n subordinate distinctions)

Section 3.1.3 (discussion of soil mottling revised; should be largely replaced by description of redoximorphic features—Appendix D).

Section 3.1.4 (additional descriptive modifiers for tubular pores added, and new discussion of micro-, meso- and macroporosity)

Section 3.1.5c (diameter break between fine and very fine roots should be 1 mm, not 0.5 mm)

Figure 3-5 (lines marking breaks from fine to medium, and coarse to very coarse for roots and pores should be the same as for structure sizes)

Section 3.1.6a (restrictive genetic horizons other that fragipan identified)

Section 3.1.6d (17 codes for restrictive layers or horizons added)

Section 3.1.6e (new reference on interpretation of cone penetrometer readings added)

Section 3.1.7f (new section of frost action potential)

Section 3.2.3 (expanded discussion of available water capacity including new SCS class definitions)

Table 3-9 (minor corrections and additions)

Section 3.3.8 (dipyridyl dye test for reduced iron added)

Table 3-11 (new table on color and occurrence of iron oxide minerals)

Section 3.2.5 (two new SCS subclasses of very slow infiltration class added)

Table 3-10 (expanded to include color criteria for placement of soil drainage classes)

Section 3.3.7 (shrink-swell indicators of clay mineralogy added)

Table 4-1 (information on several additional types of augers added)

Table 4-2 (information on additional samplers added and separated into two tables covering power-driven disturbed and undisturbed core samplers)

Appendix D (discussion of SCS's new procedures for describing redoximorphic features)

CHAPTER 1
FIELD METHODS, EQUIPMENT, AND DOCUMENTS

This chapter provides tables and checklists that can be used to help select specific field methods, and to identify equipment and documents that should be assembled prior to going into the field.

1.1 Nature of Soil Pollutants and Surface Pollution Situation

Before beginning field sampling and characterization, it is necessary to have some knowledge of the nature of soil pollutants at a site, whether they are heavy metals, toxic organics, or both, and areal extent of pollution. EPA's Environmental Sampling Expert System (ESES) defines two major types of surface pollution situations related to the areal distribution of the contaminants: (1) **large** (covers a wide area, primarily on the surface), or (2) **localized** (areas usually polluted near the source) should also be known.

Once all available site information has been evaluated, an **exploratory** soil sampling program may be undertaken to further define the nature and extent of soil pollutants, before developing a detailed soil sampling plan. Soil descriptions of near-surface soil cores (1.5 to 2 m) taken on a grid at the site and using procedures described in Appendix A.1 may provide valuable additional information at relatively low cost prior to developing either an exploratory or detailed soil sampling plan.

Existing data, or soil sampling results, will indicate the nature of pollutants. Specific contaminants of concern are broadly defined in EPA's ESES as relatively **mobile and toxic** residence time in the solid phase is relatively short, enhancing toxicity), or relatively **nonmobile and nontoxic** (residence time in the solid phase is relatively long, decreasing potential toxicity).

1.2 Soil Parameters for Field Sampling and Characterization

This field guide assumes that:

1. Physical, hydrologic, and chemical/biological parameters of soils and contaminants have been selected for description and sampling prior to the field collection phase for characterization of a contaminated site.

2. These parameters and methods or protocols are contained in a statistically sound and detailed Soil Sampling and Quality Assurance Plans for the site. Guidance for the preparation of such plans can be found in Mason (1992), Barth et al. (1989), van Ee et al. (1990), and U.S. EPA (1986, Vol. 2, Chap. 9).

Appendix C may provide assistance in selection of site and soil characterization parameters and in identification of available field, laboratory, and calculation or lookup methods for individual parameters.

This field guide is intended to assist in carrying out three major types of activities in the field:

- Description of site and soil features based on visual and tactile observation

- Field tests or measurement that involve relatively simple procedures and equipment, and

- Methods for collection of undisturbed or minimally disturbed samples for physical and microbiological characterization in the laboratory

Collection of samples for chemical characterization in the laboratory are not covered in detail in this guide, because it is assumed that these are defined in detail in the Sampling and Quality Assurance Plans for the site. However, general protocols for sample handling and preparation and for sampling with a spade and scoop, augers, or thin-wall tube samplers are contained in Appendices A.2 through A.4.

Specialized field procedures involving more complicated equipment, such as for measuring unsaturated hydraulic conductivity, are not covered in this field guide. Procedures involving such methods should be described in the Sampling Plan.

1.3 Field Description of Soils

Multiple soil profile descriptions at a site can provide a great deal of information that may be useful in evaluating the variability of soil properties, and the directions and potential for transport of contaminants in the subsurface. Detailed soil profile descriptions have not been commonly used at contaminated sites. One purpose of this field guide is to encourage greater use of this relatively easy field method.

Table 1-1 summarizes the key features that should be noted in detailed soil profile descriptions and identifies the section in this field guide that covers individual features. Preparation of a complete, detailed soil profile description requires the digging of a pit so that feature can be observed laterally as well as vertically. Although this method is time consuming, the ability to observe small-scale lateral variations in soil features associated with increased or reduced soil permeability, justifies, in most instances, a limited number of such soil profile descriptions at a site.

Where soils are not rocky, a thin-wall soil probe can be used to prepare a moderately detailed soil profile descriptions in a relatively short time per description in order to identify larger scale variations in soil characteristics. Appendix A.1 describes a general protocol for description of soil cores. Table 1-1 recommends that all the features used in preparing a pit soil profile description be observed. It should be recognized, however, that the soil core may be too small a sample of the subsurface to accurately describe a number of features, such as transitions between horizon boundaries, certain types of soil structure (columnar, for example), pore and root distribution, and genetic horizons (fragipans, for example). These features are indicated with an asterisk in Table 1-1.

Table 1-1. Suggested Soil Parameters for Field Description

Parameter	Section in Guide	Soil Profiles	Soil Samples
Horizons	3.1.1	R*	R
USDA Texture	3.1.2	R	R
Color	3.1.3	R	R
Redoximorphic Features	App D	R	
Porosity	3.1.4	R*	r
Zones of Increased Porosity/Permeability			
Soil structure	3.1.5a	R*	r
Extrastructural cracks	3.1.5b	r	r
Roots	3.1.5c	R*	r
Surface features	3.1.5d	R	
Sedimentary features	3.1.5e	R*	
Zones of Reduced Porosity/Permeability			
Genetic horizons	3.1.6a	R*	
Consistency	3.1.6b	R	
Root restricting layers	3.1.6d	R*	
Compaction	3.1.6e	r	
Moisture Condition	3.2.1	R	R
Water Table	3.2.2	R	R
Saturated Hydraulic Conductivity	3.2.4	R**	r
Clay Minerals	3.3.7	r	
Other Minerals	3.3.8	r	
Odor	3.3.2	r	

R = Recommended for all situations.

r = Recommended where climatic, geologic,and soil conditions make parameter significant.

* Soil pit may be required for accurate description of this soil feature.

** Estimation of K_{sat} class based on other observable features.

Preparation of an accurate, detailed soil profile description requires training and experience, and such descriptions are best by, or under the supervision of, someone familiar with procedures developed by the Soil Conservation Service in the U.S. Department of Agriculture.

Soil samples for chemical characterization in the laboratory should generally not double as samples for detailed soil description, because exposure to the air before placement in sample containers should be minimized. However, abbreviated descriptions should be made to identify the samples and possibly help in relating sample results to other soil profile descriptions. Table 1-1 recommends that at, a minimum, soil horizon (or depth increments), color, texture, moisture condition, and relationship to the water table be observed. The table also identifies several other features (porosity, structure, and roots) for which observations would be useful, if the nature of the sample allows (soil core) and exposure to air is less of a concern (heavy metals in aerated soil).

1.4 Field Sampling and Testing

A number of tests involving relatively simple procedures and equipment can be used to measure or characterize soil physical and chemical properties. Such tests are generally not as accurate as laboratory tests but have the advantage of being inexpensive and may be used for preliminary screening or selection of samples for more accurate laboratory testing.

Form 1-1 provides a checklist of soil engineering, physical, and chemical parameters for which field tests are described in this field guide. Special sampling procedures for microbiological characterization are covered in Section 3.3.10.

Before going into the field, this checklist should be used to identify the specific tests that appear to be of value for the site of interest. This procedure will assist in locating the appropriate section of the field guide where specific procedures are described, and in identifying equipment needs (see next section).

Form 1-1. Checklist of Soil Physical and Chemical Property Sampling and Field Test Procedures

---

Soil Physical Properties

_____Color ignition test (Section 3.1.3)
_____Extrastructural crack tests (Section 3.1.5b)
_____Fragipan identification (Section 3.1.6a)
_____Cementation test (Section 3.1.6b)
_____Bulk density (Section 3.1.6c)
_____Pocket penetrometer test (Section 3.1.6e)
_____Soil temperature regime characterization (Section 3.1.8)
_____Soil moisture (Section 3.2.1)
_____Water table estimation (Section 3.2.2)
_____Available water capacity (Section 3.2.3)
_____Saturated hydraulic conductivity class estimation (Section 3.2.4)
_____Soil drainage class placement (Section 3.2.4)

Soil Engineering Properties

_____Unified (ASTM) Texture (Section 3.1.7a)
_____Atterberg limits (Section 3.1.7b)
_____Shear strength (Section 3.1.7c)
_____Shrink-swell tests (Section 3.1.7d)
_____Corrosivity characterization (Section 3.1.7e)

Soil Chemical Properties

_____Organic matter ignition test (Section 3.3.1)
_____Cation exchange capacity/exchangeable acidity (Section 3.3.3)
_____pH (Section 3.3.4)
_____Redox potential (Section 3.3.5)
_____Electrical conductivity (Section 3.3.6)
_____Clay minerals—nitrobenzene test (Section 3.3.7)
_____Calcium carbonate—HCl test (Section 3.3.8)
_____Soluble salts—chloride and sulfate (Section 3.3.8)
_____Gypsum acetone test (Section 3.3.8)
_____Iron oxides—ignition and streak tests (Section 3.3.8)
_____Reduced iron—dipyridyl dye test (Section 3.3.8)
_____Manganese—streak and hydrogen peroxide tests (Section 3.3.8)
_____Sampling for soil microbiota (Section 3.3.10)

1.5 Field Equipment and Documents

Form 1-2 is a checklist of over 90 items that may be required for field description, analysis, and sampling of soils. Major categories covered in this checklist include: (1) documents, (2) protective equipment, (3) miscellaneous equipment, (4) site surface characterization, (5) soil description equipment/materials, (6) soil sampling equipment, (7) texture analysis and sample preparation equipment, (8) sample, equipment and waste containers and forms, and (9) field testing and analysis. Form 1-3 contains a checklist of around 20 items related to quality assurance and quality control of the sampling process, such as (1) QA/QC forms and samples, (2) material required for sample preservation and transport, and (3) decontamination equipment.

These checklists have two columns: the first is to identify those items that are needed for the site in question; the second column can be checked when the item has been obtained and packed. Required items can be identified by reviewing the site Soil Sampling Plan and Quality Assurance Plan and the checklist of field tests contained in this guide (Form 1-1).

1.6 Use of EPA's Environmental Sampling Expert System

EPA's Environmental Systems Monitoring Laboratory is developing an Environmental Sampling Expert System (ESES) which ultimately will integrate a Geographic Information System and Site Description System (including data quality objectives, quality assurance and quality control, and site description) with a Knowledge Frame Manager (for analysis interpretation of data and report preparation) for contaminated sites.

The **Guide to Site and Soil Description for Hazardous Waste Site Characterization** (Cameron, 1991), provides the basis for the site and soil components for metals in the ESES. Site and soil parameters (called Object/Attributes in the ESES), are assigned "values" which have significance for contaminant transport and fate.

Form 1-2. Soil Description/Sampling Equipment and Documents Checklist

---

Check first column to identify needed items. Check second column when item has been obtained and packed prior to leaving for the field.

Documents

___ ___ Sampling plan
___ ___ Quality assurance plan
___ ___ Health and safety plan
___ ___ Log books

Protective Equipment

___ ___ Protective suits
___ ___ Boots
___ ___ Gloves (inner/outer)
___ ___ Duct tape
___ ___ Respirators, respirator cartridges, and/or dust masks
___ ___ Raingear and/or warm clothing
___ ___ Insect repellent (should not contain chemicals that will be target analytes or in matrix spikes)

Miscellaneous

___ ___ Keys for access to site, graphite lubricant for locks
___ ___ Folding table
___ ___ Camera and film
___ ___ Flashlight and extra batteries
___ ___ Toolbox, including hacksaw
___ ___ Calculator

Site Surface Characterization

___ ___ Max/min thermometer
___ ___ Humidity gage or sling psychrometer
___ ___ Hand-held anemometer
___ ___ 4-foot staffs with flags or flagging (for wind direction indicators)
___ ___ Clinometer for slope measurement
___ ___ 6-foot rod and colored tape to mark eye level for clinometer readings   (Continued)

Form 1-2. (Continued)

Soil Description Equipment/Materials

___ ___ Field notebook, pencils, ballpoints, and markers
___ ___ Clipboard with cover
___ ___ Microcassette recorder, spare microcassettes and batteries
(optional for notetaking)
___ ___ Map of soil sampling locations
___ ___ Soil profile and related information forms (Form 3-1)
___ ___ Unified (ASTM) texture determination form (Form 3-2)
___ ___ Other soil data collection forms:

___ ___ Carpenters rule (for measuring horizon depth) and/or steel tape
___ ___ 30-cm by 2-m plastic sheet for placing soil cores for description
___ ___ Munsell Soil Color book
___ ___ Knife (for cleaning exposed soil surfaces)
___ ___ Nails (for marking horizon boundaries)
___ ___ 10 power hand lens (surface features, mineral identification, carbonate test)
___ ___ Sand size and coarse fragment determination scales
___ ___ 1/2 in. mesh (for estimating areal distribution of features on excavated profile)
___ ___ Tile probe (soil depth determination in rocky soil)
___ ___ Stiff, 2-mm wire for crack depth measurement
___ ___ Graded sand of uniform color for crack characterization (excavation method)
___ ___ Stereoscopic microscope (5 to 6 in. working distance, 20 to 80 power)
___ ___ Small high intensity 6v flexible lamp (for illuminating microscope)

Soil Sampling Equipment (specified in Sampling Plan)

Hand-held sampling devices
(Check all items that may be required at the site.)
___ ___ Shovel/spade
___ ___ Spoons
___ ___ Scoops
___ ___ Screw-type auger   (Continued)

1-9

Form 1-2. (Continued)

___ ___ Barrel/bucket auger (regular, sand, mud, stone, planer, in-situ soil recovery)

___ ___ Thin-wall tube

___ ___ Chisel rock breakers

___ ___ Crescent wrenches, vice grips, pipe wrenches (for changing drill rod length and sampling tips)

___ ___ Weighted plastic mallet

___ ___ Tube sampler cleaning tool

Power driven sampling devices
(Check types planned for use at the site)

___ ___ Auger

___ ___ Split spoon

___ ___ Thin-wall tube samplers

Texture Analysis and Sample Preparation Equipment

___ ___ Sieves (3 in., 1/2 in., No. 10, for characterizing coarse fraction)

___ ___ Hanging spring scale with canvas sling or pail (for weighing coarse fragments)

___ ___ No. 10 mesh stainless steel screen (for TOC and semivolatile samples)

___ ___ No. 10 mesh Teflon® screen (for metals samples)

___ ___ Compositing bucket/mixing containers (stainless steel, glass, Teflon®-lined suitable for all contaminants; Al pans for any contaminant except Al; plastic for metals analysis only)

___ ___ 1-m square piece of suitable plastic, canvas or rubber sheeting (for sample preparation)

Sample, Equipment, and Waste Containers and Forms

___ ___ Brown plastic trash bags for dirty equipment

___ ___ White plastic trash bags for clean equipment

___ ___ Ziplock type plastic bags for protecting equipment that cannot be decontaminated (cameras, notebooks, etc.)

___ ___ Sample containers (sealed clean and labeled plus 20 percent)

___ ___ Plastic bags for sample containers

___ ___ Sample description/identification forms

___ ___ Sample labels/tags

___ ___ Soil moisture tins   (Continued)

Form 1-2. (Continued)

Field Testing and Analysis

___ ___ Photoionization meter (PID) or flame ionization detector (FID)
___ ___ Calibration gases for meters
___ ___ Hydrogen gas for FID
___ ___ Specialty gas meters (HCN, etc.)
___ ___ Explosimeter
___ ___ Scale/balance (0.1 gram accuracy) for weighing of samples
(moisture, bulk density, organic matter)
___ ___ Infrared lamp, or small oven, and thermometer scaled to at least
120°C (for drying samples for moisture, bulk density, and
organic matter tests)
___ ___ Portable gas soldering torch and porcelain crucible or small tin
(not Al) with wire bracket or tongs (for ignition tests)
___ ___ Saran-ketone mixture (bulk density clod method) or sand-
measuring or rubber balloon apparatus (bulk density excavation
method)
___ ___ pH measurement kit and standard solutions (spare batteries, if
necessary), and/or color dyes, pH test strips
___ ___ Glass or plastic stirring rod (pH test)
___ ___ Small containers for mixing water and soil (pH, specific
conductance tests)
___ ___ Conductivity meter and specific conductance standards
___ ___ Quart container of distilled deionized water in squeeze bottle (for
pH, texture, and carbonate tests)
___ ___ Common laboratory spatulas (for texture tests)
___ ___ Porcelain spot plate (carbonate test, iron oxide, manganese oxide
scratch tests)
___ ___ Clean glass rod (carbonate test)
___ ___ 10-percent HCl in plastic squeeze bottle (carbonate test)
___ ___ Solution of malachite green in nitrobenzene (for clay minerals
test)
___ ___ Hydrogen peroxide (manganese test, organic matter tests)
___ ___ Neutral solution of α α'-dipyridyl dye in 1 N ammonium acetate
(reduced iron test)
___ ___ Test tubes or plastic vials and 5-percent silver nitrate, and 5-
percent barium chloride solutions (chloride and sulfate tests)
___ ___ Small stoppable bottle, filter paper, and acetone (gypsum test)
___ ___ Field sampling glove box and core paring tool (for aseptic core
samples for microbiological analysis--see Section 3.3.10)

DESCRIPTION AND SAMPLING: A FIELD GUIDE

Form 1-3. Soil Sampling Quality Assurance/Quality Control Checklist

---

Check first column to identify needed items. Check second column when item has been obtained and packed prior to leaving for the field.

Forms

___ ___ List of sample locations where duplicates and other QA samples are to be taken
___ ___ Sample alteration form (Form 3-1), multiple copies
___ ___ Field audit checklist (Form 3-2)
___ ___ Soil sample corrective action form (Form 3-3)

QA Samples (check types specified in QA Plan)*

Double-Blind Samples

___ ___ Field evaluation samples (FES)
___ ___ Low level field evaluation samples (LLFES)
___ ___ External laboratory evaluation samples (ELES)
___ ___ Low level external laboratory evaluation samples (LLELES)
___ ___ Field matrix spike (FMS)
___ ___ Field duplicate (FD)
___ ___ Preparation split (PS)

Single-Blind Samples

___ ___ Field rinsate blanks (FRB)—also called field blanks, decontamination blanks, equipment blanks, and dynamic blanks
___ ___ Preparation rinsate blank (PRB)—also called sample bank blanks
___ ___ Trip blank (TB)—also called field blank

Sample Preservation and Transport

___ ___ Chest or 6-pack cooler
___ ___ Ice
___ ___ Max/Min thermometer
___ ___ Chain-of-custody forms and seals
___ ___ Shipping forms
___ ___ Analytical analysis request forms, if different from chain-of-custody forms   (Continued)

1-12

Form 1-3. (Continued)

Decontamination

___ ___ Decontamination vessel
___ ___ Wash solution(s)—should be specified in Sampling Plan.
___ ___ Garden spray cans for wash fluids
___ ___ Rinse solutions (acetone, deionized water)
___ ___ Labels for containerized wastes (solid or liquid)

* See van Ee et al. (1990) for more detailed discussion of these types of samples.

Form 3-1 (Soil Profile and Related Information) contains space to record observations related to all the site and soil knowledge frames in EPA's ESES. Data on this form can be coded on Forms 1-4 and 1-5 (Coding Sheet for Use of ESES Site and Soil Knowledge Frames) to aid in data interpretation using the ESES or Cameron (1991). These forms also indicate the Section in this field guide that covers description procedures and nomenclature. Where standard SCS descriptive procedures do not readily allow assignment of a "value" for the ESES, definitions used in the ESES are provided for use during field observations. Cameron (1991) provides additional information on definitions of terms.

Form 1-4. Coding Sheet for ESES Site Knowledge Frames

Object/Attribute	Source*	Value
Nature of Heavy Metal Pollutants (enter elements with excessive concentrations— see Table 3-1 in Cameron (1991)	Test results	Mobile/ Nonmobile/ Toxic Nontoxic ___ ___ ___ ___ ___ ___ ___ ___ ___ ___ ___ ___
Climate/Weather	Lookup	___ Humid ___ Temperate ___ Dry
Macrofauna and Mesofauna	2.6	___ Many ___ Common ___ Few
Slope	2.2	___ Steep (>12%) ___ Moderate (3-12%) ___ Flat (0-3%)
Surface Erosion/ Erodibility	2.3	___ Severe ___ Moderate ___ Slight to none
Surface Pollution Situations	1.1	___ Large areas ___ Localized areas
Surface Runoff	2.4	___ Rapid (H, VH) ___ Medium (M) ___ Slow (L, VL) ___ Ponded (N)
Vegetation	2.5	___ Dense ___ Scattered/Sparse ___ Absent
Wind Speed/ Direction	2.1.2	___ Gale ___ Breezy ___ Calm to light

* Use observations recorded on Form 3-1 or refer to indicated section number in this guide. Appendix C provides information on laboratory and lookup methods.

Form 1-5. Coding Sheet for ESES Soil Knowledge Frames

Object/Attribute	Source*	Value
Bulk Density (g/cc)	3.1.6c Lab	___ Low (<1.3) ___ Medium (1.3-1.6) ___ High (>1.6)
Cation Exchange Capacity (meq/100 g soil)	3.3.3 Lab	___ Low (<12) ___ Medium (12-20) ___ High (>20)
Clay Minerals	3.1.2 3.3.7 Lab	___ Abundant (>27%) ___ Mod/Slight(1-27%) ___ None/Neg. (<1%)
Color	3.1.3	___ Dark ___ Red and Yellow ___ Brown ___ Gray/Whitish ___ Mottled
Compaction	3.1.6e	___ High ___ Moderate ___ Low/Slight
Consistency	3.1.6b	___ High ___ Moderate ___ Low/Weak ___ Cemented
Corrosivity	3.1.7e Lookup	___ High ___ Moderate ___ Low
Electrical Conductivity (Salinity, mmhos/cm)	3.3.6 Lab	___ Nonsaline (<2) ___ Slight (2-4) ___ Moderate (4-8) ___ Very (8-16) ___ Extremely (>16)
Fertility Potential	3.3.9 Lab	___ High ___ Moderate ___ Low (Continued)

Form 1-5. (Continued)

Object/Attribute	Source*	Value
Horizons	3.1.1	___ Master Horizons ___ Transitional ___ Disturbed ___ Buried
Hydraulic Conductivity (μm/sec)	3.2.4	___ High (>10) ___ Moderate (0.1-10) ___ Low (0.01-0.1) ___ Inhibited (<0.01)
Infiltration/Percolation (cm/hr)	3.2.5	___ High (>5) ___ Medium (1.5-5.0) ___ Low (0.15-1.5) ___ Inhibited (<0.15)
Microbiota	3.3.10 Lab	___ Abundant ___ Common ___ Few ___ None
Moisture Conditions	3.2.1 Lab	___ Wet ___ Moist ___ Dry
Odor	3.3.2	___ High ___ Mod/Slight ___ None
Organic Matter	3.3.1 Lab	___ Abundant ($>4\%$) ___ Moderate (2-4%) ___ Sparse ($<2\%$)
Porosity	3.1.4	___ Coarse (>5mm) ___ Medium (2-5mm) ___ Fine (0.5-2mm) ___ Very Fine (<0.5mm)
Reaction (pH)	3.3.4	___ Acid (<6.6) ___ Neutral (6.6-7.3) ___ Alkaline (>7.3)

(Continued)

Form 1-5. (Continued)

Object/Attribute	Source*	Value
Redox Potential	3.3.5	___ High Oxidized ___ Intermediate ___ Highly Reduced
Roots	3.1.5c	___ Many ___ Common ___ Few
Structure Grades	3.1.5a	___ Structureless ___ Weak ___ Moderate ___ Strong
Surface Features	3.1.5d	___ Prominent ___ Distinct ___ Faint
Temperature	3.1.8	___ High ___ Medium ___ Low
Temperature Regimes	3.1.8 Lookup	___ Pergelic ___ Cryic ___ Frigid-Isofrigid ___ Mesic-Isomesic ___ Thermic-Isothermic ___ Hyperthermic- Isohyperthermic
Texture Classes	3.1.2	___ Fragmental ___ Sandy ___ Silty ___ Loamy ___ Clayey ___ Organic Soils

* Use observations recorded on Form 3-1 or refer to indicated section number in this guide. Appendix C provides information on laboratory and lookup methods.

CHAPTER 2
SITE CHARACTERISTICS

This chapter covers a number of weather-related factors that affect the ease or difficulty of soil description and sampling in the field (Section 2.1) and other site surface features at locations where soils are described or sampled. These features include: slope (Section 2.2), surface erosion (Section 2.3), surface runoff (Section 2.4), vegetation (Section 2.5) and macro- and mesofauna associated with the surface and subsurface (Section 2.6).

2.1 Climate and Weather

Climate exerts a profound influence on soil directly through soil forming and weathering processes such as precipitation, evapotranspiration, and temperature, and indirectly by its influence on vegetation. Climate is a known factor at a site that is usually evaluated by analysis of meteorologic records from nearby weather stations, although in some instances detailed monitoring of meteorologic parameters such as precipitation, temperature, and wind may be required as a part of site characterization and remediation. This guide does not cover methods for systematic monitoring of climatic factors.

Weather refers to the state of the atmosphere at a site during field investigation activities. Unusual weather conditions are usually noted during the sampling and description of soils. Weather usually doesn't become a concern during field work unless conditions, such as rain or snow, adversely affect the carrying out of field procedures or create health and safety concerns for field personnel. The major weather parameters to be monitored during field work include air temperature, wind speed and direction, and humidity. The site health and safety officer is primarily responsible for evaluating adverse weather conditions and pacing field activities accordingly, but field personnel should communicate their own feelings about working under adverse weather conditions.

2.1.1 Air Temperature

Air temperature is primarily a concern when it is at one extreme or the other. Heat becomes a special concern when protective clothing must be worn on site, because the impervious material used increases sweating and the possibility of heat stress from dehydration. Extreme cold makes sampling and notetaking difficult. Max-min thermometers are relatively inexpensive and daily extremes should be recorded along with periodic observations through the day, if appropriate. Wind speed (see Section 2.1.2) should be monitored when working in the winter to determine wind chill temperatures. Humidity (see Section 2.1.3) should be monitored when the climate is humid and temperatures are high.

2.1.2 Wind Speed and Direction

In winter, wind speed should be monitored to estimate wind chill temperature. Hand-held anemometers can be used for this purpose. High winds create unfavorable conditions for soil sampling, especially when soil is dry, because of the possibility of contamination from blowing surface soil and the mobilization of contaminated subsoil that is brought to the surface. When it is breezy, personnel should position themselves upwind during soil sampling, and avoid sampling in locations where contaminated soil might blow into the exclusion area. Four-foot staffs with flags or flagging placed around the site can serve as wind direction indicators. Wind speed and direction should be recorded at each location where soils are sampled.

EPA's Environmental Sampling Expert System (ESES) defines wind speed classes as follows:

Gale: >32 mph (>37 knots)

Breezy: 4 to 32 mph (3-37 knots)

Calm to light: <4 mph (<3 knots)

2.1.3 Humidity

Relative humidity, the ratio of measured atmospheric water vapor pressure to that which would prevail under saturated conditions, is the most commonly used measure of atmospheric moisture. For general field use, relative humidity can be measured using a humidity gage, but, if very accurate measurements are desired, a sling psychrometer should be used.

Humidity is primarily a concern for field operations when it is very high or very low. Very high humidity associated with high temperatures increases the danger of heat stress in field personnel, especially when protective clothing must be used on site. When sampling for soil moisture when the humidity is low, special care should be taken to minimize exposure of soil to the air to avoid drying before the sample is sealed.

2.2 Slope

Slope is an important site feature that influences the distribution of precipitation between the soil and surface runoff, and the movement of soil water. Slope gradient is usually measured as a percentage, but may be measured in degrees. Both gradient and the length of the slope (to the point where surface runoff loses its energy and deposits suspended soil particles) are required for estimating erosion using the Universal Soil Loss Equation (see Section 2.3 below). Slope shape and topographic position influence the movement of water on the surface and in the subsurface. Slope aspect affects the moisture status of soil, with southern exposures usually drier than northern exposures due to increased evapotranspiration.

Soil surveys prepared by the U.S. Soil Conservation Service (SCS) differentiate soil map units in upland areas by the dominant soil series and a slope range (such as 0-2 percent, 2-6 percent, 6-12 percent, etc.) that is based on soil management considerations. Slope classes are based on slope gradient limits as follows (Soil Survey Staff, 1993):

Classes		Slope Gradient Limits (%)	
Simple Slopes	Complex Slopes	Lower	Upper
Nearly level	Nearly level	0	3
Gently sloping	Undulating	1	8
Strongly sloping	Rolling	4	16
Moderately steep	Hilly	10	30
Steep	Steep	20	60
Very steep	Very steep	>45	

Different SCS county soil surveys may specify different slope ranges for a slope class within the lower and upper limits identified above.

The following slope features should be observed when preparing a soil description:

Gradient (percent or degrees)—Measured using a clinometer and a rod with a marking at the observers eye level. Siting through the clinometer up or downslope to the marker on the rod allows a direct reading in percent or degrees. Accurate readings require that (1) the line of siting is perpendicular to the contour of the slope, and (2) that there is no change in slope gradient over the distance the siting is taken.

Length (if erosion potential is evaluated).

Shape—Convex, concave, or flat.

Topographic position—Summit, shoulder, backslope, footslope, toeslope, or floodplain.

2.3 Surface Erosion and Erodibility

Field evaluation of surface erosion has two components: (1) assessment of soil loss or deposition that has occurred in the past, and (2) evaluation of the future erosion potential.

In upland soils, the amount of erosion can be inferred by comparing observed texture and color in the A and B horizons (see Section 3.1.1 for definitions of horizons) with a nearby undisturbed soil in a similar topographic setting (if available), or with a soil profile description prepared by the SCS for the soil series to which the soil belongs. The thickness of the A horizon is reduced in **moderately** eroded soils, and may show mixing with the B horizon if the soil has been cultivated. Rill erosion, the removal of soil through the cutting of many small, but conspicuous channels, may be evident on unvegetated soil. In **severely** eroded soils, most of the topsoil is missing and gully erosion (channels that cannot be obliterated by ordinary tillage) may be evident. Soils with **slight** to no erosion have fully developed A horizons and surface material showing little evidence of erosion.

In depressional areas, the thickness of soil that has accumulated as a result of accelerated erosion can be measured by finding the top of the natural A horizon, provided the eroded material can be differentiated by color, texture, and other soil features. Surface contaminated soil preferentially concentrates in such areas, and special sampling may be desirable.

Use of the Universal Soil Loss Equation (USLE) or its revised version (RUSLE) to estimate erosion potential from a site requires the following field observations: (1) slope gradient and length (see Section 2.2 above) and (2) vegetation (Section 2.5). The soil erodibility factor (K) can usually be obtained from SCS soil series interpretation sheets. If classification of a soil is uncertain, the K factor can be estimated using a soil erodibility nomograph (SCS, 1993—Exhibit 618-10). The following soil properties must be described or estimated to use these soil nomographs: (1) percent silt plus very fine sand, (2) percent sand (0.10 to 2.0 mm), (3) percent organic matter (see Section 3.3.1), (4) soil structure (see Section 3.1.5a), and (5) permeability class (see

Section 3.2.4. If a soil is rocky (>5 percent by volume rock fragments >2 mm), the K factor needs to be further adjusted. SCS (1983, 1992) provide more detailed information on estimating K and adjusting for rock fragments.

EPA's ESES defines soil erodibility classes based on estimated annual soil loss, using the USLE or RUSLE, as follows:

Severe: >10 metric tons/hectare.

Moderate: 2.5 to 10 metric tons/hectare.

Slight: <2.5 metric tons/hectare.

2.4 Surface Runoff

Surface runoff potential is important for evaluating the potential for transport of contaminants at the soil surface to surface streams or water bodies.

SCS defines six runoff classes that can be used for qualitative comparison of runoff from different locations at a site (See Table 2-1). Placement requires measurement of the slope gradient (see Section 2.2) and measurement or estimation of the saturated hydraulic conductivity (see Section 3.2.4).

Computation of runoff by SCS's Curve Number Method requires placement of soils in hydrologic groups. If the hydrologic class of the soil of interest is not known, it can be estimated based on saturated hydraulic conductivity (K_{sat}—see Section 3.2.4), and water table characteristics (Section 3.2.2) using Table 2-2. Langan and Lammers (1991) proposed more detailed criteria for placement of soils in hydrologic groups based on soil data collected in the eastern U.S. as part of U.S. EPA's Direct/Delayed Response Project. Table 2-3 presents these more detailed criteria, which use drainage class (Table 3-10), with adjustments for depth to root limiting/impermeable layers (Sections 3.1.6a and 3.1.6d), texture class in stratified soils (Section 3.1.2), and relationship between perched water, impermeable layers, and slope.

Table 2-1. Index Surface Runoff Classes

Slope Gradient (%)	VH	H	Runoff Classes* K_{sat} Class** MH	ML	L	VL
Concave***	N	N	N	N	N	N
<1	N	N	N	L	M	H
1-5	N	VL	L	M	H	VH
5-10	VL	L	M	H	VH	VH
10-20	VL	L	M	H	VH	VH
>20	L	M	H	VH	VH	VH

* Abbreviations: Negligible-N; very low-VL; low-L; medium-M; high-H; and very high-VH. These classes are relative and not quantitative.

** See Section 3.2.4 for definitions. Assumes that the lowest value for the soil occurs at <0.5 m. If the lowest value occurs at 0.5 to 1 m, reduce runoff by one class. If it occurs at >1 m, then use the lowest saturated hydraulic conductivity <1m. VL K_{sat} is assumed for soils with seasonal shallow or very shallow free water.

*** Areas from which no or very little water escapes by flow over the ground surface.

Table 2-2. Criteria for Placement of Hydrologic Soil Groups

Soil Group*	Criteria* K_{sat} (μm/s)**	Free Water Depth/Duration
A	>55	>1.5 m
B	5.5 to 55	>1 m
C	.55 to 5.5	>0.5 m
D	<.55	0 to 0.5 m; transitory through permanent***

* The criteria are guidelines only. They are based on the assumption that the minimum saturated hydraulic conductivity occurs within the uppermost 0.5 m. If the minimum occurs between 0.5 and 1 m, then K_{sat} for the purpose of placement is increased one class. If the minimum occurs below 1 m, then the value for the soil is based on values above 1 m using the rules previously given.

** Values are midpoints of the four highest K_{sat} classes in Table 3-9.

*** See Section 3.2.2 for lengths of durations.

Adapted from Soil Survey Staff (1993)

Table 2-3 Expanded Criteria for Hydrologic Soil Groups

Soil Group	Criteria

A   **Drainage classes***: well drained, somewhat excessively, and excessively drained.
Root limiting/impermeable layers**: > 100 cm, except when (1) paralithic contact is rapidly permeable, paralithic contact can be < 100 cm (e.g., weathered granodiorite high in biotite); (2) folists can have bedrock as shallow as 25 cm.
Stratified soils***: sandy or sandy-skeletal.
Perched water table: well to excessively drained soils may shift to Group B as indicated below.

B   **Drainage classes***: moderately well drained (except when root limiting layer/impermeable layer is 50 to 100 cm).
Root limiting/impermeable layers**: > 100 cm (except folists which can have bedrock at < 25 cm).
Stratified soils***: Loamy or loamy-skeletal.
Perched water table: *Lithic or paralithic contact 50 to 100 cm or other root limiting layer/very slowly permeable material between 75 and 100 cm*: well to excessively drained with rapid permeability on slopes > 3%; well to excessively drained with moderate permeability on slopes > 15%. *Other root-limiting layer or slowly or very slowly permeable layer between 50 to 75 cm*: well to excessively drained with rapid permeability on slopes > 8%; well to excessively drained with moderately rapid permeability on slopes > 15%; well to excessively drained with moderate permeability on slopes > 25%. *Any slope when root-limiting/impermeable layer is between 100 to 150 cm as follows*: well drained with moderately rapid to very slow permeability; moderately well drained with moderately rapid to moderately slow permeability.

C   **Drainage classes***: somewhat poorly drained (except when root limiting layer/impermeable layer is 50 to 100 cm).

(Continued)

Table 2-3 (Continued)

Soil Group	Criteria
	Root limiting/impermeable layers**: 50 to 100 cm (moderately well drained soils); >100 (somewhat poorly drained soils). **Stratified soils*****: clayey or clayey-skeletal >50 cm. **Perched water table**: any slope when root-limiting/impermeable layer is between 100 to 150 cm when somewhat poorly drained soils have moderately rapid to moderately slow permeability.
D	**Drainage classes***: poorly and very poorly drained. **Root limiting/impermeable layers****: <50 cm; paralithic contact <50 cm and lithic contact <100 cm; 50 to 100 cm (somewhat poorly drained soils) **Stratified soils*****: clayey or clayey-skeletal <50 cm. **Perched water table**: Any slope when root limiting/impermeable layer is <50 cm.
A/D, B/D C/D	Soils classified by SCS as dual hydrologic groups are placed in the highest class when fully drained by artificial measures and intermediate classes if partially drained (A/D and B/D).

Adapted from Langan and Lammers (1991)

* See Table 3-10 for definitions

** includes root limiting genetic soil horizons such as fragipans, duripans, lithc and paralithic contacts (see Sections 3.1.6a) and 3.1.6d), slowly permeable clay layers, and dense glacial tills. Impermeable layer = slowly or very slow permeability (see Section 3.2.5).

** Weighted particle-size class of all horizons between 25 and 100 cm (see Section 3.1.2). Soils meeting the taxonomic definition of strongly contrasting particle-size classes are placed based on the most slowly permeable texture, regardless of their relative position.

2.5 Vegetation

Vegetation at a site serves as an indicator of site history and site productivity and is a major determinant in erosion potential at a site (see Section 2.3).

Features observed and noted should include the nature, kind, extent, and distribution of plants and plant cover. The charts in Figure 3-3 (Section 3.1.2) for estimating areal percentages of coarse fragments and mottles also can be used to estimate amount of vegetative cover. Vogel (1987) describes more precise methods for measuring vegetation cover such as the point-quadrant method, rated microplots, and line intercepts.

Stunted vegetation or discolored leaves may be an indication of toxic effects from contaminants in the soil. In heavy metal contaminated sites, sampling of vegetation along with soil may be desirable to assess exposure through bioaccumulation.

EPA's ESES defines qualitative vegetation classes as follows:

Dense: Site completely covered with vegetation of predominant forms or varying composition and species, usually with slow temporal variability.

Scattered to sparse: Plant cover, aerial or soil surface vegetation, is intermittent or infrequent at site.

Absent: No visible macrovegetation can be observed, but some scattered soil vegetation cover (e.g., algal-lichen crusts or mosses) may be evident.

2.6 Macro- and Mesofauna

Soil macrofauna, such as burrowing animals, earthworms, and larger insects that can be measured in centimeters, and mesofauna, such as smaller mollusks and arthropods, affect soil profiles by mixing, changing, and moving soil material. The activities of soil macro- and mesofauna tend to increase the secondary permeability of soil horizons and thus provide preferential paths for subsurface migrations of contaminants.

Surface features that animals produce include termite mounds, ant hills, heaps of excavated earth beside burrows, the openings of burrows, paths, feeding grounds, and earthworm or other castings.

These features can be described in terms of (1) number of structures per unit ares, (2) proportionate area occupied, and (3) volume of above ground structures. **Krotovinas**, irregular tubular streaks of soil material, with contrasting color or texture, resulting from filling of tunnels made by burrowing animals, may be observed in soil pits.

Where soils are contaminated with heavy metals, collection of soil macro- or mesofauna for chemical analysis may provide evidence of bioaccumulation for exposure assessment, provided that similar species on nearby uncontaminated soils with similar characteristics can be obtained for comparison.

Macro- and mesofauna are described for EPA's ESES by species and abundance per unit area as follows:

Class	Number/m^2
Macrofauna	
Many	> 10
Common	5-10
Few	1-5
None	0
Mesofauna	
Many	> 1,000,000
Common	100-1,000,000
Few	10-100
None	< 10

CHAPTER 3
FIELD DESCRIPTION AND ANALYSIS OF SOILS

This chapter presents field test and soil description procedures that can be done visually or with simple field equipment. Appendix C identifies references where more detailed information can be found about more complex field procedures for measuring specific soil parameters.

For nonengineering applications, the soil taxonomy of the U.S. Soil Conservation Service is the most widely used system for describing and classifying soils. The basic reference on this system is **Soil Taxonomy**, Agricultural Handbook (AH) 436 (Soil Survey Staff, 1975). Revisions and amendments to AH 436 are periodically published in looseleaf form as the **National Soil Taxonomy Handbook** (issue No. 16 was released August 4, 1992). The most recent edition of the pocket-sized **Keys to Soil Taxonomy** (Soil Survey Staff, 1992) is the best concise reference source on classification of soils, and is recommended for use in the field (price and ordering information are given in the reference section). The introduction to each new edition identifies significant changes in soil description and classification procedures that have occurred since the last edition. SCS occasionally issues **Soil Taxonomy Notes** to clarify, interpret and add consistency to the application. Additional guidance for the field description and classification of soils comes from the **National Soil Survey Center Technical Note Handbook** (Note No. 3 was issued July 1, 1992).

The most up-to-date comprehensive reference for SCS soil description procedures, is a major revision of Agricultural Handbook 18, **Soil Survey Manual** (Soil Survey Staff, 1993). Most of the soil description procedures in this guide are taken from the final manuscript the new manual (page proofs were being checked by SCS at the time this second edition was completed). The **National Soils Handbook** (Soil Conservation Service, 1983), focussing on procedures for soil mapping, and interpretation of soils information, is another important reference source. Part 603.02 (Estimated Soil Properties and Features) is especially useful in providing guidance of ways to interpret and estimate soil physical, chemical and hydrologic properties. In 1992 a revised and expanded draft of Part 603.02 was released as Part 618 (Interpretations-Soil Properties and Qualities) in the draft **National Soil Survey Interpretations Handbook** (SCS, 1992). Other important references for field investigation procedures for noncontaminant parameters are published by SCS (1971 and 1984). The reference section provides some additional information on the SCS publications discussed above. All of the above documents are available for inspection at any state office of the Soil Conservation Service, and may be available in area offices or district offices where SCS soil scientists are stationed.

Procedures from the geologic literature for description of unconsolidated material below the weathering zone are also included in this chapter.

3.1 Soil Physical Parameters

A number of soil physical parameters such as coarse fragments, pores, mottled colors, roots, lateral features, and mineral concentrations require area measurements on the ground surface or the wall of a pit for conversion to volume or weight percentages. Section 3.1.2 includes charts for estimating proportions of a feature at the surface, but if more accurate measurements are desired, traverses on an arbitrary grid can be used.

Coarse screening such as hardware cloth or rat wire with a 1/2-inch mesh makes a convenient and durable grid for small and medium objects as large as 2 to 4 inches (5 to 10 cm). A screen 1 foot square is suitable for most situations. Marking every fifth, tenth or twentieth wire in each direction, or at intersections, with paint makes counting easier. When the screen is tacked over the area to be measured, a small wire pointer pushed into the soil at the intersection of each wire allows the most accurate counting of features in the grid. SCS (1971, Section I2.7) provides additional guidance on making linear and volume measurements.

Form 3-1 provides a sample form for description of a soil profile. This form follows the sequence of features for description of soil horizons in this chapter. Standard forms used by SCS and coding sheets for computer programs, which automatically prepare narrative soil profile descriptions, can also be used.

Form 3-1. Soil Profile and Related Information

Soil Type or Designation _____

Date _____ ID No._____

Described by _____

Location _____

_____ Elevation _____

Wind Speed/Direction (2.1.1) _____

Other Weather Conditions (2.1) _____

Parent Material _____

Topographic Position (2.2)

       Slope Gradient _____ Slope Length _____

       Slope Shape _____ Slope Aspect _____

Erosion (2.3) _____

Surface Runoff Class (2.4) _____

Vegetation (2.5) _____

Macro- and Mesofauna (2.6) _____

Bulk Density (3.1.6c) _____ Compaction (3.1.6e) _____

Engineering Properties (3.1.7):

 USCS Texture _____ Shrink-Swell _____

 Shear Strength _____ Corrosivity _____

Soil Temperature/Regime (3.1.8) _____

Water Table (3.3.3)

 Depth (Max/Min) _____ Thickness, if Perched _____

 Duration _____ Drainage Class (3.2.4) _____

Infiltration (3.2.5) _____

CEC (3.3.3) _____ Redox Potential (3.3.5) _____

Electrical Conductivity (3.3.6) _____

Fertility Potential (3.3.9) _____ Microbiota (3.3.10) _____

Soil Classification _____

Additional Notes _____

Form 3-1. (Continued)

Horizon (3.1.1) _____ _____ _____
  Depth _____ _____ _____
  Boundary _____ _____ _____
Texture (3.1.2)
  Fines (<2 mm) _____ _____ _____
  >2 mm (%) _____ _____ _____
  >2 mm (description) _____ _____ _____
Matrix Color (3.1.3)
  Moist/Dry _____ _____ _____
Redox Concentrations (App. D)
  Fe Masses _____ _____ _____
  Fe/Mn Concretions _____ _____ _____
  Pore Linings _____ _____ _____
Redox Depletions (App. D)
  In Matrix _____ _____ _____
  Along Pores _____ _____ _____
Clay Minerals (3.3.7) _____ _____ _____
Other Minerals (3.3.8)
  HCl (carbonates) _____ _____ _____
  H_2O_2 (Mn oxides) _____ _____ _____
  Dipyr. dye (FeII) _____ _____ _____
  Other _____ _____ _____
Pores (3.1.4) _____ _____ _____
Structure (3.1.5a) _____ _____ _____
Roots (3.1.5c) _____ _____ _____
Surface/Sedimentary
  Features (3.1.5d, e) _____ _____ _____
Consistency (3.1.6b)
  Moist/Dry _____ _____ _____
  Cementation _____ _____ _____
Water State (3.2.1) _____ _____ _____
Ksat (3.2.4) _____ _____ _____
Odor (3.3.2) _____ _____ _____
pH (3.3.4) _____ _____ _____

A dug pit that provides a lateral view of soil horizons is best for accurate and detailed soil profile description. Thin-wall tube samplers are the next best alternative. Appendix A.1 outlines procedures for soil descriptions using tube samplers and augers. Comprehensive descriptions using one or two pits, along with less detailed tube/auger sampler descriptions at each sampling site, would probably provide the maximum amount of useful data.

3.1.1 Soil Horizons

Table 3-1 provides a key to SCS's 1981 system for designating **master horizons**, **layers**, and **transitional horizons**, along with the lower case letters that are used for subordinate distinctions within horizons. If an SCS soil survey of the site is available, the soil series descriptions should be reviewed for a general idea of the types of horizons likely to be encountered. SCS soil surveys published prior to around 1984 contain soil profile descriptions using the 1962 system. Table 3-2 compares the 1962 and 1981 systems and provides approximate equivalencies where nomenclature has changed.

In glaciated areas, it may be useful to make more precise designations for the C horizon. Table 3-3 shows subdivisions and diagnostic characteristics of four types of C horizons recognized by the Illinois State Geological Survey. If this notation is used in the description, it should be clearly noted on the field sheet.

Key features to record are the depth and characteristics of the boundary between horizons. The following notation can be used to describe horizon boundaries or contacts:

Distinctness	Topography
a — abrupt (<2 cm)	s — smooth (nearly a plain)
c — clear (2-5 cm)	c — clear (pockets w/ width > depth)
g — gradual (5-15 cm)	i — irregular (pockets w/ depth > width)
d — diffuse (>5 cm)	b — broken (discontinuous)

Table 3-1. Definitions and Designation Nomenclature for USDA Soil Horizons and Layers (Adapted from SSSA, 1987)

Master Horizons and Layers

O Horizons—Layers dominated by organic material, except limnic layers that are organic.

A Horizons—Mineral horizons that formed at the surface or below an O horizon and (1) are characterized by an accumulation of humified organic matter intimately mixed with the mineral fraction and not dominated by properties characteristic of E or B horizons; or (2) have properties resulting from cultivation, pasturing, or similar kinds of disturbance.

E Horizons—Mineral horizons in which the main feature is loss of silicate clay, iron, aluminum, or some combination of these, leaving a concentration of sand and silt particles of quartz or other resistant materials.

B Horizons—Horizons that formed below an A, E, or O horizon and are dominated by (1) carbonates, gypsum, or silica, alone or in combination; (2) evidence of removal of carbonates; (3) concentrations of sesquioxides; (4) alterations that form silicate clay; (5) formation of granular, blocky, or prismatic structure; or (6) a combination of these.

C Horizons—Horizons or layers, excluding hard bedrock, that are little affected by pedogenic processes and lack properties of O, A, E, or B horizons. Most are mineral layers, but limnic layers, whether organic or inorganic are included.

R Layers—Hard bedrock including granite, basalt, quartzite and indurated limestone or sandstone that is sufficiently coherent to make hand digging impractical.

Transitional Horizons

Two kinds of transitional horizons occur. In one, the properties of an overlying or underlying horizon are superimposed on properties of the other horizon throughout the transition zone (i.e., AB, BC, etc.). In the other, distinct parts that are characteristic of one master horizon are recognizable and enclose parts characteristic of a second recognizable master horizon (i.e., E/B, B/E, and B/C).

(continued)

Table 3-1. (Continued)

Alphabetical Designation of Horizons

Capital letters designate master horizons (see definitions above).
Lowercase letters are used as suffixes to indicate specific characteristics of the master horizon (see definitions below). The lowercase letter immediately follows the capital letter designation.

Numeric Designation of Horizons

Arabic numerals used as (1) suffixes to indicate vertical subdivisions within a horizon and (2) prefixes to indicate discontinuities.

Prime Symbol

The prime symbol (') is used to identify the lower of two horizons having identical letter designations that are separated by a horizon of a different kind. If three horizons have identical designations, a double prime ('') is used to indicate the lowest.

Subordinate Distinctions within Horizons and Layers

a — Highly decomposed organic material where rubbed fiber content averages < 1/6 of the volume.

b — Identifiable buried genetic horizons in a mineral soil.

c — Concretions or hard nonconcretionary nodules of iron, aluminum, manganese, or titanium cement.

e — Organic material of intermediate decomposition in which rubbed fiber content is 1/6 to 2/5 of the volume.

f — Frozen soil in which the horizon or layer contains permanent ice.

g — Strong gleying in which iron has been reduced and removed during soil formation or in which iron has been preserved in a reduced state because of saturation with stagnant water.

h — Illuvial accumulation of organic matter in the form of amorphous, dispersible organic matter-sesquioxide complexes, where sesquioxides are in very small quantities and the value and chroma of the horizons are < 3.

i — Slightly decomposed organic material in which rubbed fiber content is more than about 2/5 of the volume.

(continued)

Table 3-1. (Continued)

---

k — Accumulation of pedogenic carbonates, commonly calcium carbonate (NSSC Technical Note No. 1, September 16, 1991, provides additional guidance on use of this suffix).

m — Continuous or nearly continuous cementation or induration of the soil matrix by carbonates (km), silica (qm), iron (sm), gypsum (ym), carbonates and silica (kqm), or salts more soluble than gypsum (zm).

n — Accumulation of sodium on the exchange complex sufficient to yield a morphological appearance of a natric horizon (NSSC Technical Note No. 3, July 1, 1992, provides additional guidance on use of this suffix).

o — Residual accumulation of sesquioxides.

p — Plowing or other disturbance of the surface lay- ers by cultivation, pasturing, or similar uses.

q — Accumulation of secondary silica.

r — Weathered or soft bedrock including saprolite; partly consolidated soft sandstone, siltstone, or shale; or dense till that roots penetrate only along joint planes and are sufficiently incoherent to permit hand digging with a spade.

s — Illuvial accumulation of sesquioxides and organic matter in the form of illuvial, amorphous dispersible organic matter-sesquioxide complexes, if **both** organic matter and sesquioxides components are significant and the value and chroma of the horizon are >3.

t — Accumulation of silicate clay that either has formed in the horizon and is subsequently translocated or has been moved into it by illuviation.

v — Plinthite which is composed of iron-rich, humus-poor, reddish material that is firm or very firm when moist and that hardens irreversibly when exposed to the atmosphere under repeated wetting and drying.

w — Development of color or structure in a horizon but with little or no apparent illuvial accumulation of materials.

x — Fragic or fragipan characteristics that result in genetically developed firmness, brittleness, or high bulk density.

y — Accumulation of gypsum.

z — Accumulation of salts more soluble than gypsum.

Table 3-2. Comparison of the 1962 and 1981 USDA Soil Horizon Designation Systems

---

Alphabetical Designation of Horizons

Capital letters designate master horizons in **both** systems, but there are some changes in specific letter designations (see below).

Lowercase letters are used as suffixes to indicate specific characteristics of the master horizon in both systems, but there are some changes in specific letter designations (see below). In the 1981 System, the lowercase letter always immediately follows the capital letter designation.

Numeric Designation of Horizons

1962 System: Arabic numerals used as suffixes to (1) indicate kind of O, A, or B horizon, and (2) indicate vertical subdivisions of a horizon; roman numerals used as prefixes to indicate lithologic discontinuities.

1981 System: Arabic numerals used as suffixes to indicate vertical subdivisions within a horizon **and** as prefixes to indicate discontinuities. Their use to indicate kind of O, A, or B horizon has been eliminated.

Prime Symbol

1962 System: The prime used to identify the lower sequum of a soil having two sequa (horizon sequences), although not for a buried soil.

1981 System: The prime used to identify the lower of two horizons having identical letter designations that are separated by a horizon of a different kind. If three horizons have identical designations, a double prime is used on the lowest.

(continued)

3-9

Table 3-2 (Continued)

Comparisons of Horizon Designations (see Table 3-1 for definitions)

Master Horizons		Subordinate Horizon Distinctions	
1962	1981	1962	1981
O	O	—	a
O1	Oi,Oe	b	b
O2	Oa,Oe	cn	c
A	A	—	e
A1	A	f	f
A2	E	g	g
A3	AB or EB	h	h
AB	—	—	i
A&B	E/B	ca	k
AC	AC	m	m
B	B	sa	n
B1	BA or BE	—	o
B&A	B/E	p	p
B2	B or Bw	si	q
B3	BC or CB	r	r
C	C	ir	s
R	R	t	t
		—	v
IIB23t	2Bt3	—	w
		x	x
		cs	y
		sa	z

Source: Adapted from Guthrie and Witty (1982).

Table 3-3. Subdivisions of the C Horizon Used in Illinois

Horizon	Mineralogy	Carbonates	Color	Structure
C1	Strongly altered	Leached	Uniform, mottled or stained	Some soil structure, peds with clay films; structure of parent material—blocky; layered, or massive—dominant; often porous
C2	Altered	Unleached	Uniform, mottled, or stained	Less soil structure, clay films along joints; structure of parent material—blocky; layered, or massive—dominant; often porous
C3	Partly altered	Unleached	Uniform; rare stains	Massive, layered, or very large blocky; conchoidal fractures; dense
C4	Unaltered	Unleached	Uniform	Massive or layered, conchoidal fractures; dense

Source: Follmer et al. (1979).

A **disturbed** soil has been truncated or manipulated to the extent that its principle pedogenic characteristics have been severely altered or can no longer be recognized. A **buried** soil, or paleosol, is covered by an alluvial, loessial (windblown), or other depositional surface mantle, and usually lies below the weathering profile of the soil at the land surface. As noted in Section 3.1.5, buried soils may have high secondary porosity compared to materials above and below it, forming a potential zone for preferential movement of contaminants.

3.1.2 Soil Texture Classes

Soil texture, the relative proportions of silt-, sand-, and clay-sized particles (also called particle- or grain-size distribution), is an important property from which many other soil parameters can be estimated or inferred. This section focuses on the USDA soil texture classification system. Many other classification systems have been developed, but of these, only the ASTM (Unified) system, which is oriented toward soil engineering applications, is covered in this field guide (see Section 3.1.7a).

Figure 3-1 shows a modified USDA soil texture triangle. Classification is based on the fine fraction (less than 2 mm), with modifiers applied where coarse fragments are more than 15 percent by volume. For example, consider a sample where the <2 mm fraction plots on the texture triangle as a sand, and contains 40 percent rock fragments which are mostly around 30 mm in diameter. Looking in the upper right corner of Table 3-4, the adjective modifier for 35 to 60 percent coarse fragments is the dominant rock plus the word "very". Looking in the middle of Table 3-4, the SCS adjective for coarse fragments around 30 mm would be gravelly or coarse gravelly. Thus, the full texture description would be "very coarse gravelly sand". The noun used to describe the coarse fragments would be pebbles or coarse pebbles. Table 3-4 also shows simplified descriptors for the >2 mm fraction based on the Wentworth scale, which is more commonly used by geologists.

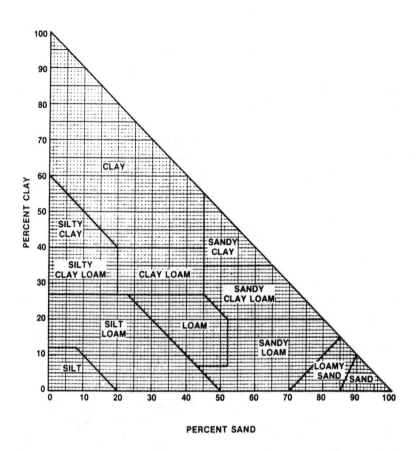

Percentages based on < 2mm fraction. Example of Use: A soil material with 15% clay and 60% sand is a sandy loam. If more than 15% of the soil material is larger than 2.0 mm, the texture term includes a modifier (Table 3-4). Example: gravelly sandy loam.

Figure 3-1. USDA modified soil texture triangle. (Gee and Bauder, 1986; copyright Soil Science Society America, used by permission).

3-13

Table 3-4. Abbreviations and Designations for USDA Soil Texture Classes (including coarse fraction)

<2 mm Fraction	>2 mm Fraction (SCS)
(See texture triangle, Figure 3-1)	**Adjective modifier** (see text explanation)
	<15% none
s - sand	15-35% dominant rock
ls - loamy sand	35-60% dominant rock + very
sl - sandy loam	>60% (>5% fines) dominant
l - loam	rock + extremely
si - silt	>60% (<5% fines) dominant
sil - silt loam	rock adjective
cl - clay loam	**Other Descriptive Features of**
sicl - silty clay	**>2 mm fraction**
loam	Percent
sc - sandy clay	Roundness
sic - silty clay	Mineralogy/rock type
c - clay	Sorting

Rock Descriptors for > 2 mm Fraction (SCS)

Adjective/Noun	Shape/Size Rounded, subrounded angular, or irregular (diameter, mm)
g - gravelly/pebbles	2-76
fg - fine gravelly/fine pebbles	2-5
mg - medium gravelly/medium pebbles	5-20
cg - coarse gravelly/coarse pebbles	20-76
k - cobbly/cobbles	76-250
st - stony/stones	250-600
b - bouldery/boulders	>600
	Flat (long, mm)
st - stony/stones	280-600
b - bouldery/boulders	>600
ch - channery/channers	2-150
flg - flaggy/flagstones	150-380

Table 3-4. (Continued)

Simplified Descriptors for > 2 mm Fraction (IDEM)

gn - granules	2-4
pb - pebbles	4-64
fpb - fine pebbles	4-16
cpb - coarse pebbles	16-64
cb - cobbles	64-256
b - boulders	>256

Source: Soil Survey Staff (1991) and IDEM (1988).

Sandy soil classes are often divided into subclasses according the coarseness of the sand grains. Figure 3-2 shows the criteria for subclasses of sandy soils. Table 3-4 summarizes abbreviations and designations for recording USDA soil texture in the field. Figure 3-3 provides charts for estimating percentages of coarse fragments in a soil horizon. Field determinations based on estimated percentages of clay and sand should be verified by laboratory analysis of samples.

The following general groupings of texture classes are sometimes used (see Table 3-4 for abbreviations):

Sandy (light or coarse) — s, ls
Silty (medium) — si, sil, sicl
Loamy (medium) — sl, l, cl, scl (sil, si, sicl may be included in this category)
Clayey (heavy or fine) — sc, sic, c

The USDA soil taxonomy defines particle-size classes for differentiation of the soils at the family level (Soil Survey Staff 1975, 1992) as described below:

Fragmental—90 percent or more (by volume) rock fragments and voids; less than 10 percent (by volume) particles less than 2.0 mm in diameter.

Basic soil class	Subclass	Soil separates				
		Very coarse sand, 2.0-1.0 mm.	Coarse sand, 1.0-0.5 mm.	Medium sand, 0.5-0.25 mm.	Fine sand, 0.25-0.1mm.	Very fine sand, 0.1-0.05 mm.
Sands	Coarse sand	25% or more			Less than 50%	Less than 50%	Less than 50%
	Sand	25% or more			Less than 50%	Less than 50%
	Fine sand	Less than 25%		–or–	50% or more	Less than 50%
	Very fine sand					50% or more
Loamy Sands	Loamy coarse sand	25% or more			Less than 50%	Less than 50%	Less than 50%
	Loamy sand	25% or more			Less than 50%	Less than 50%
	Loamy fine sand	Less than 25%		–or–	50% or more	Less than 50%
	Loamy very fine sand					50% or more
Sandy Loams	Coarse sandy loam	25% or more			Less than 50%	Less than 50%	Less than 50%
	Sandy loam	30% or more		–and–			
		Less than 25%			Less than 30%	Less than 30%
	Fine sandy loam	–or– Between 15 and 30%			30% or more	Less than 30%
	Very fine sandy loam	–or– Less than 15%			More than 40%	30% or more

*Half of fine sand and very fine sand must be very fine sand.

Figure 3-2. Percentage of sand sizes in subclasses of sand, loamy sand, and sany loam basic texture classes.

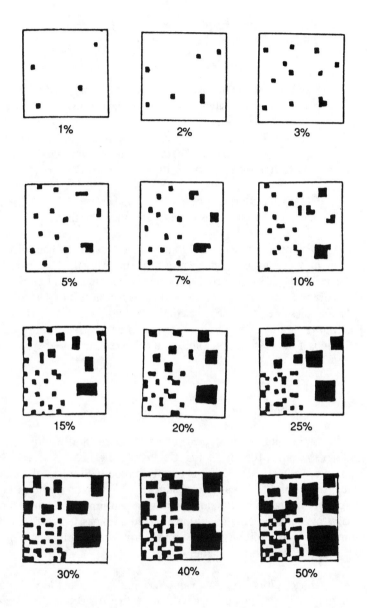

Figure 3-3. Charts for estimating proportions of coarse fragments and mottles (each fourth of any one square has the same amount of black).

Skeletal—Rock fragments make up 35 percent or more by volume. The dominant fine earth fraction (sandy, loamy, or clayey) is used as a modifier.

Sandy—Texture of fine earth is sand or loamy sand with < 50 percent very fine sand; clay < 35 percent; rocks < 35 percent.

Loamy—Texture of fine earth is very fine sand or finer; clay < 35 percent; rocks < 35 percent. Subdivision include coarse-loamy, fine-loamy, coarse-silty, and fine-silty.

Clayey—Texture of fine earth is > 35 percent clay; rocks < 35 percent. Subdivisions include fine and very fine.

Different particle-size class terminology is used when soils are formed in volcanic ash and ejecta. The fragmental class is called **pumiceous** if > 2/3 of the coarse fragments are pumice and **cindery** if < 2/3 of the coarse fragments are pumice. Fine earth fractions of such soils are classified as **ashy, medial,** or **hydrous. Keys to Soil Taxonomy** (Soil Survey Staff, 1992) should be referred to for precise definitions if the presence of Andisols at a site requires use these terms.

3.1.3 Soil Color

Soil color is described using Munsell Soil Color charts (available from Munsell Color Company, 2441 N. Calvert St., Baltimore, MD 21218). Color is usually a good indicator of the redox status (see Section 3.3.5 and Appendix D) of a horizon (uniform high chroma colors indicate oxidizing conditions; uniform low chroma colors usually indicate reducing conditions; mixed chromas indicate variable saturation). In addition, low chroma, low value colors are often indicative of high organic matter content.

Dark colors have low value (generally < 3) and low chroma (generally < 2). **Red** colors generally have hues of 10R or 2.5YR, and values and chromas > 3. **Yellow** colors generally have values and chromas > 6 and hues of 7.5YR, 10YR, or 2.5Y. **Brown** colors typically have values from 3 to 6, chroma from 3 to 5, and hues of 7.5YR or 10YR. **Gray or whitish** colors may be of any hue with chroma <2 and values generally >3. The Munsell Soil Color charts provide precise descriptors for any soil color reading.

The following data should be recorded in field description of color:

Color name.
Color notation (chroma, hue, value).
Water state (*moist* or dry).
Physical state (*broken* through ped is the standard state. Others are: rubbed between fingers (moist), or crushed/crushed and smoothed (dry).
Soil Mottling is usually an indication of variable saturation, and is described according to abundance, size, and color contrast. Figure 3-3 provides charts that may help in estimating mottle abundance, and Figure 3-4 provides guidance in identifying contrast.

Abbreviations, descriptors, and criteria for description of mottles are:

Abundance

f — few (<2%)
c — common (2-20%)
m — many (>20%)

Size

1 — fine (<5 mm)
2 — medium (5-15 mm)
3 — coarse (>15 mm)

Contrast (see Figure 3-4)

f — faint (1 or 2 units H,C,V)
d — distinct (2-4 units H,C,V)
p — prominent (4-5 units H,C,V)

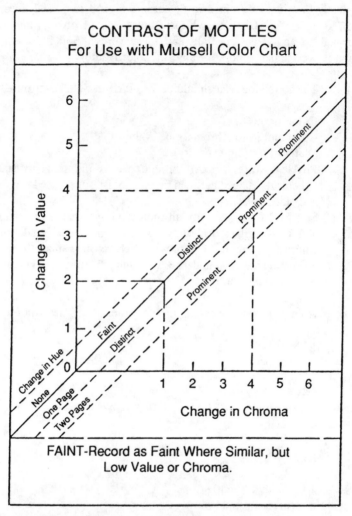

Figure 3-4. Guide for designation of mottle/redoximorphic feature contrast.

Shape (spots, streaks, bands, tongues, tubes)
Location (inped, exped)
Boundaries (sharp — like a knife edge; clear — color grade over <2 mm; diffuse — color grade over >2 mm)

Important Note on Description of Soil Mottles: Redoximorphic features now replace soil mottles in SCS soil description procedures for aquic soil conditions. Appendix D describes in more detail how these features are described and the relationship between redoximorphic features and mottles. The new procedures for description of redoximorphic features are recommended for accurate characterization of aquic conditions in soils. The above descriptors may still be useful for abbreviated descriptions from cores and can help in the interpretation of soil profile descriptions from SCS soil surveys published prior to 1993.

Color Ignition Test

An ignition test using 2 or 3 grams of soil can provide useful information for interpretation of natural soil colors. The equipment and procedures described in Section 3.3.1 for organic matter should be used, except that weighing of the samples is not required. The following provides some guidance for interpreting colors:

Organic matter contributes black, brown, reddish (spodic horizons), and grayish colors. It burns away leaving a whitish residue if it is the only colored material.

Minerals such as quartz, which make up the bulk of sand and silt-sized particles, are mostly colorless or pale colored to gray. These particles will not change color with ignition. Mineral grains may be cemented with lime or silica or stained with iron oxide, especially in dry regions. SCS (1971, Section I7.1) describes procedures for cleaning mineral surfaces of cement and stains.

Iron oxides are red, brown, or yellow. If browns and yellows become redder and brighter with ignition, highly hydrated iron oxide (goethite) is present. **Ferrous** (reduced) iron is indicated by gray, blue, or green colors, and turns red when ignited to form hematite (Section

3.5.8 also includes a field chemical test for ferrous iron). Table 3-11 provides additional information on interpretation of colors of iron oxide minerals.

Manganese oxides form black and purple bodies and effervesce vigorously in a 5 percent solution of hydrogen peroxide (Section 3.3.8). Dark reddish brown and dark brown surface soils in the southeastern U.S. usually contain enough manganese oxides to give a positive reaction to peroxide.

Soil Color, a recent publication by the Soil Science Society of America edited by Bingham and Ciolkosz (1993), contains additional useful information on the intepretation of soil color.

3.1.4 Soil Porosity

Laboratory analysis is required for accurate determination of soil porosity, but field description of soil pores can provide useful qualitative data for estimating permeability and characterization of soil variability at a site. Johnson et al. (1960) provide more detailed guidance on classification and description of soil pores.

SCS describes pores according to (1) abundance, (2) size, (3) distribution within the horizon, and (4) type. Figure 3-5 can be used to estimate pore size in the field. Below are abbreviations, descriptors and criteria for describing pores in the field:

Abundance Classes	No./Unit Area
1 — few	< 1
2 — common	1-5
3 — many	> 5

Size Classes	Diameter	Unit Area
vf — very fine	< 0.5 mm	1 cm^2
f — fine	0.5-2 mm	1 cm^2
m — medium	2-5 mm	10 cm^2
cos — coarse	5-10 mm	10 cm^2
vcos — very coarse	> 10 mm	1 m^2

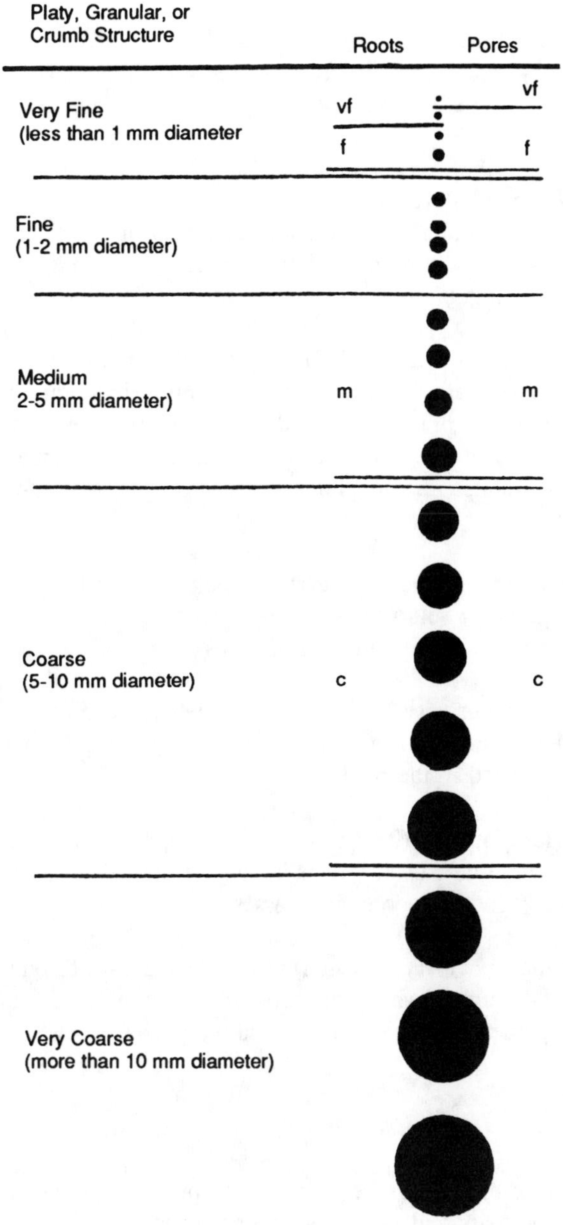

Figure 3-5. Charts for estimating pore and root size.

Distribution Within Horizons
in — inped (most roots and pores are within peds)
ex — exped (most roots and pores follow interfaces between peds)

Types of Pores
v — vesicular (approximately spherical or elliptical)
t — tubular (approximately cylindrical and elongated)
i — irregular

Johnson et al. (1960) suggest the following additional descriptive modifiers for tubular pores: **simple** (individual pores are single tubules, not branched); **dendritic** (individual pores branch like plant roots); **continuous** (individual pores extend through the horizon); and **discontinuous** (individual pores extend only part-way through the horizon).

Luxmoore (1981) suggested classifying soil pores into three categories based on soil water dynamics, which may prove useful when interpreting field descriptions of soil pores:

Macroporosity (<0.01 mm) where capillary flow is dominant. This type of porosity is too small to be easily described in the field).

Mesoporosity (0.01 to 1 mm) where gravitational flow occurs in unsaturated soils. This is roughly equivalent to the fine and very fine SCS pore size classes.

Macroporosity (>1 mm) where channel flow through the soil is dominant when the soil is ponded or saturated. This includes medium, coarse and very coarse SCS pore size classes.

3.1.5 Zones of Increased Porosity/Permeability

Weathering and other soil-forming processes often increase the secondary porosity and permeability of unconsolidated materials. In addition, the mode of deposition of unweathered materials may create vertical and lateral variations in permeability that should be described. Increased secondary porosity is usually confined to the zone of soil weathering near the surface. Buried soils (paleosols) in glaciated areas represent zones of potential lateral movement of contaminants due to increased secondary porosity, if underlain by less permeable material.

3.1.5a Soil Structure Grades

Soil structure is an important feature that affects the movement of contaminants in soil. Contaminants often move preferentially along the interfaces between soil structure units. SCS describes soil structure according to shape (see Figure 3-6 for illustrations), grade, and size (see Figure 3-7 for charts). Below are abbreviations, descriptors, and criteria for describing soil structure in the field:

Grade

0 — structureless (massive or single grain)
1 — weak (poorly defined individual peds)
2 — moderate (well formed peds, but not distinct)
3 — strong (durable peds, quite evident in place; will stand displacement)

Size	Shape		
	pl-platy gr-granular cr-crumb	abk-angular blocky sbk-subangular blocky	cl-columnar pr-prismatic
vf—very fine	< 1 mm	< 5 mm	< 10 mm
f—fine	1-2 mm	5-10 mm	10-20 mm
m—medium	2-5 mm	10-20 mm	20-50 mm
c—coarse	5-10 mm2	0-50 mm	50-100 mm
vc—very coarse	> 10 mm	> 50 mm	> 100 mm

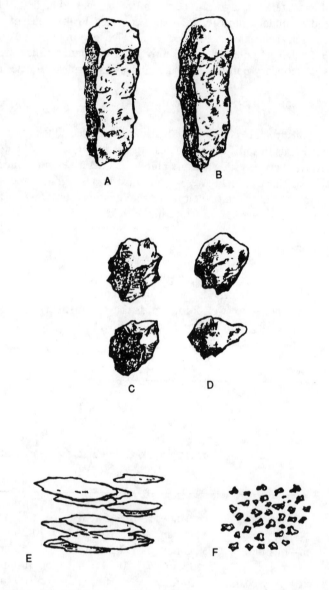

Figure 3-6. Drawings illustrating some of the types of soil structure: A, prismatic; B, columnar; C, angular blocky; D, subangular blocky; E, platy; and F, granular.

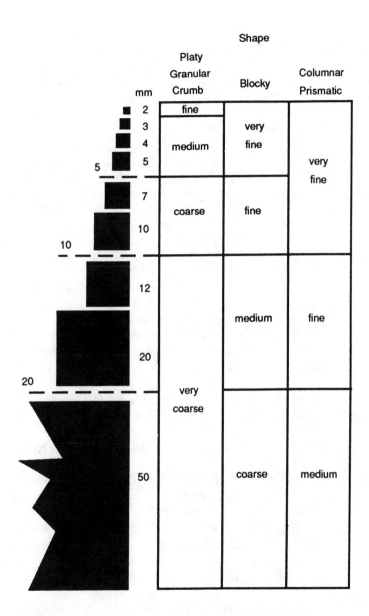

Figure 3-7. Charts for estimating size class of different structural units (see also Figure 3-5 for platy, granular, and crumb structures).

Accurate identification columnar or prismatic structure generally requires a soil pit. Blocky structure can usually be identified in cores taken from thin-wall samplers, but size class cannot always be accurately identified. Augers disturb the soil too much to allow accurate description of soil structure.

3.1.5b Extrastructural Cracks

Cracks are macroscopic vertical planar voids with a width much smaller than length and depth which result from soil drying. Extrastructural cracks extend beyond the planar surfaces between soil structural units and represent major potential channels for increasing infiltration of water during precipitation and preferential contaminant movement into the soil. The presence of irreversible cracks in the soil increases ponded infiltration (see Section 3.2.5).

SCS define four major types of extra-structural cracks:

Surface-initiated reversible cracks form as a result of drying from the surface downward. They close after relatively slight surficial wetting and have little influence on ponded infiltration rates.

Surface-initiated irreversible cracks form from freeze-thaw action and other processes, and do not close completely when rewet. They extend to the depth that frost action has occurred, and act to increase the ponded infiltration rates.

Subsurface-initiated reversible cracks form in subsoils with a high shrink-swell potential as the soil dries from field capacity. They close in a matter of days if the horizon becomes moderately moist or wet. They extend to the surface (unless the surface horizon does not permit the propagation of cracks), and increase ponded infiltration and rates of soil evaporation.

Subsurface-initiated irreversible cracks are permanently present in the subsurface.

Tests for Crack Characterization

A crack index value may be obtained by using a blunt wire, approximately 2 mm in diameter. More detailed characterization of cracks can be accomplished by pouring loose sand into the crack and excavating after wetting and after the crack has closed.

Penetrant cracks are 15 cm or more in depth as measured by an inserted wire. **Surface-connected** cracks occur at the ground surface or are covered by up to 10 to 15 cm of soil material that has very high or high saturated hydraulic conductivity with soft, very friable or loose consistency. Surface connected cracks increase ponded infiltration.

Crack development is primarily associated with clayey soils and is most pronounced in high shrink-swell soils (see Section 3.1.7d).

3.1.5c Roots

The penetration of plant roots into the soil increases the secondary porosity of soil, and, after the plant dies and the root decomposes, leaves channels for more rapid flow of water through the soil. The absence of roots in the near surface is also an indication of reduced porosity/permeability.

A soil pit is required to accurately describe soil roots. SCS conventions for describing roots in soil are similar to those for soil pore description although, the criteria for abundance and size classes are slightly different from those for pores. These classes are described below:

Abundance Classes	No./Unit Area
1 — few	< 1
v1 — very few	< 0.2
m1 — moderately few	0.2-1
2 — common	1-5
3 — many	> 5

Size Classes	Diameter	Unit Area
vf — very fine	<1 mm	1 cm^2
f — fine	1-2 mm	1 cm^2
m — medium	2-5 mm	10 cm^2
cos — coarse	5-10 mm	10 cm^2
vcos — very coarse	>10 mm	1 m^2

Distribution Within Horizons

in — inped (most roots are within peds)
ex — exped (most roots follow interfaces between peds)

Figure 3-5 provides charts for estimating root size.

3.1.5d Surface Features

Surface or lateral features such as clay films and silt coatings are often indicators of areas of increased permeability in the soil. Stress formations, on the other hand, are good indicators of active shrinking and swelling of clays in the soil. SCS describes lateral features according to (1) kind, (2) amount, (3) distinctness, and (4) location. Section I3.1 in SCS (1971) provides more detailed guidance on the observation of features on ped faces. A stereoscopic microscope illuminated by a high-intensity lamp is useful for more detailed observation of lateral features. Criteria for descriptors of lateral features are as follows:

Kind	Amount
Coatings	vf — very few (<5%)
Clay films	f — few (5-25%)
Clay bridges	c — common (25-50%)
Sand	m — many (>50%)
Silt	
Other	
Stress formations	
Pressure faces	
Slickensides	

Distinctness

Faint — Requires 10 power magnification; little contrast in properties with material.

Distinct — Detectable without magnification although magnifier or other tests may be required for positive identification. Contrast in properties with adjacent material evident.

Prominent — Conspicuous without magnifier; properties contrast sharply with adjacent material.

Location

Surfaces of peds, channels, pores, primary particles or grains, soil fragments, rock fragments, nodules, or concretions.

3.1.5e Sedimentary Features

Sedimentary features in unweathered unconsolidated materials often allow inferences to be made about the depositional history of the material and, when combined with particle-size distribution data, assist in the location of zones of more rapid lateral movement of contaminants. The main features described are type and orientation. The following types and orientations are taken from the Indiana Department of Environmental Management guidelines for description of unconsolidated material at hazardous waste sites (IDEM, 1988):

Type (describe thickness)	Orientation
1 — bedding/lamination	ver — vertical
2 — cross-stratification	hor — horizontal
3 — deformation in bedding	obl — oblique
4 — bedding surface structures	
5 — fossils/bioturbation	
6 — massive (no structure)	

3.1.6 Zones of Reduced Porosity/Permeability

Low-permeability soil or other horizons in unconsolidated material inhibit the downward movement of contaminants. Ground water tends to perch above such zones and transport contaminants laterally. Soil features indicating such zones include (1) slowly permeable genetic horizons, (2) very firm and very hard consistency classes, (3) high bulk density, (4) root restricting layers, and (5) high penetration resistance.

3.1.6a Genetic Horizons

Fragipans and cemented or indurated horizons formed by the precipitation of carbonates (petrocalcic horizons), silica (duripans), iron oxides (petroferric horizon or plinthite), or other minerals (petrogypsic) are distinctive features of certain soil series. Determination of degree of cementation requires wetting the sample for at least an hour (see 3.1.6b).

Fragipans, which occur widely in areas of the eastern United States, present substantial problems in the field identification. Witty and Knox (1989) suggest the following essential characteristics of a fragipan:

1. One or more of the subhorizons are brittle at or near field capacity throughout the subhorizon or at least in any large prismatic structural units that have horizontal dimensions of 10 cm or more and constitute 60 percent or more of the volume.

2. Air-dried fragments of 5 to 10 cm in size from any part of the horizon slake or fracture when placed in water.

3. Roots are virtually absent, except in vertical streaks, between any large prismatic structural units that have horizontal dimensions of 10 cm or more and that constitute 60 percent or more of the volume.

4. There is evidence of pedogenesis in the form of mottles, clay films, or vertical streaks that define large prisms.

Many or all fragipans also exhibit one or more of the following characteristics:

1. Relatively low vertical saturated hydraulic conductivity (slow or very slow permeability), as measured or as revealed by evidence of perched water in the form of mottles, an E horizon, or seasonal seepage immediately above the pan.

2. High bulk density (commonly <35 percent porosity of the fine-earth fraction) relative to overlying horizons.

3. Large prisms defined by vertical streaks that are arranged in a polygonal pattern on a horizontal exposure.

Both vertical and horizontal exposure of the horizon is required for complete description of a fragipan. More comprehensive and definitive criteria for identification of fragipans have been proposed and are under review within SCS (Glocker and Quandt, 1992). Cemented horizons, such as duripans and petrocalcic horizons are generally found in well-developed soils in arid and semi-arid regions. Section 3.1.6d (Restrictive Layers) provides a list of genetic and other types of horizons that restrict downward movement of contaminants.

3.1.6b Rupture Resistance (Consistency)

Rupture resistance, also called consistence or consistency, is a readily observed feature in the field. Terms used to describe rupture resistance vary depending on the moisture content of the sample tested (see Table 3-5). The very firm (moist) and very hard (dry) classes, and those that are firmer or harder, are indicative of reduced porosity/permeability. The footnotes to Table 3-5 describe the procedure for estimating rupture resistance classes. SCS recommends practice with scales to gain a more precise tactile sense of the transition between the different classes.

Table 3-5. Rupture Resistance (Consistency) Classes

Moist (≥DS*)	Dry (DM,DV*)	Conditions of Failure**	Stress Applied***
Loose	Loose	Specimen not obtainable	—
Very friable	Soft	Fails under **very slight** force applied slowly between thumb and forefinger	<8N
Friable	Slightly hard	Fails under **slight** force applied slowly between thumb and forefinger	8-20N
Firm	Hard	Fails under **moderate** force applied slowly between thumb and forefinger	20-40N
Very firm	"	Fails under **strong** force applied slowly between thumb and forefinger	40-80N
Extremely firm	Very hard	Cannot be failed between thumb and forefinger but can be by applying pressure slowly with hands; fails if placed on a hard surface and gentle force applied underfoot	80-160N
"	Extremely hard	Cannot be failed in hand, but can be crushed or broken underfoot by the full body weight applied slowly	160-800N
Rigid	Rigid	Cannot be failed underfoot by full body weight but can be by <3J blow	800N-3J
Very rigid	Very rigid	Cannot be failed by blow of 3J	>3J

Footnotes for Table 3-5

*See Table 3-8 for definitions of abbreviations.

**Standard specimens should be block-like and 25 to 30 mm on edge. If specimens smaller than the standard size must be used, corrections should be made for class estimates (i.e., a 10-cm block will require about one-third the force to rupture as will a 30-cm block). Stress is applied along the vertical in-place axis of the specimen by compressing it between extended thumb and forefinger, between both hands, or between the foot and a nonresilient flat surface. If the specimen resists compression, a 1-kg weight is dropped from progressively greater heights up to 30 cm, until rupture.

***Both force (newtons; N) and energy (joules; J) are employed. One newton is equivalent to the force necessary to acclerate a 1-kg mass one meter per second per second. One joule is the energy delivered by dropping a 1-kg weight 10 cm. A tactile sense of class limits may be learned by applying pressure through the tips of the fingers or ball of the foot to postal or bathroom scales.

Source: Adapted from Soil Survey Staff (1993).

Cementation Test. Degree of cementation can be estimated by applying the rupture resistance tests in Table 3-5 to an air-dried specimen that has then been placed in water for at least an hour. Terms used to described cementation are as follows (use Table 3-5 to estimate stress applied):

Class	Stress Applied
Uncemented	$<8N$
Weakly cemented	8-80N
Moderately cemented	80-800N
Strongly cemented	800N-3J
Indurated	$>3J$

Other consistence classes used in the field are plasticity and stickiness. These are a function of clay content and are covered in Section 3.1.7a.

EPA's Environmental Sampling Expert System (ESES) defines the following consistency classes:

3-35

High: A soil when wet that shows high cohesion of soil particles, or adhesion of soil particles to other substances.

Moderate: A soil when moist that shows moderate cohesion or adhesion.

Low to Weak: A soil, usually dry, that shows reduced or poor cohesion or adhesion.

Cemented: A type of soil that remains hard or brittle after an air-dried specimen has been placed in water for at least 1 hour.

3.1.6c Bulk Density

Accurate measurement of bulk density, expressed as g/cc, requires weighing a known volume of soil or determining both the weight and volume of an undisturbed soil sample. Some commonly used methods are described below.

Core Method. This method involves collecting a core of a known volume using a thin-wall sampler (to minimize disturbance of the soil sample), and transporting the core to the laboratory for weighing. Inserting a sampling cylinder inside the sampling tube allows other measurements to be made in the laboratory such as pore-size distribution, hydraulic conductivity, and water retention. Core samples should be placed in moisture-proof containers to maintain field moisture content.

Hole or Excavation Method. Bulk density can be determined directly in the field by excavating a quantity of soil, drying and weighing it (in the field or laboratory), and determining the volume of the excavation. This volume can be determined by measuring the volume of sand required to fill the excavation, or placing a rubber balloon in the excavation and measuring the amount of water or some other liquid required to fill the space. SCS (1971) and Blake and Hartge (1986) describe equipment and field procedures for these methods.

Coated-Clod Method. Bulk density of cohesive soils can be measured by coating a clod with a saran-ketone mixture and comparing the weights of the clod in air and water. Blake and Hartge (1986) describe this procedure in detail. This procedure can be done in a field laboratory with a scale that weighs accurately to 1 gram in a range of 500 to 1,000 grams. SCS (1971) describes procedures for determining the field-moist density and dry density of a clod. With these measurements one can determine the minimum and maximum density, field moisture capacity, percentage of volume change, and ratio of moist and dry volumes for calculation of the coefficient of linear extensibility (COLE—see Section 3.1.7d).

The bulk density at which resistance to root penetration is high varies with texture as follows (SCS, 1983):

Family Texture	Bulk Density
Sandy	> 1.85
Coarse-Loamy	> 1.80
Fine-Loamy	> 1.70
Coarse-Silty	> 1.60
Fine-Silty	> 1.50
Fine	> 1.35

3.1.6d Restrictive Layers

A restrictive layer is a nearly continuous layer that has one or more physical or chemical properties which significantly reduces soil water and air permeability or increases excavation difficulty. Restrictive layers are an indicator that contaminants will tend to move laterally along the top of the restrictive layer rather than vertically in the soil profile.

SCS classifies root restricting soil layers, whether they are genetic horizons (Section 3.1.6a) or not, based on a combination of other soil properties as follows (the appropriate section where specific features are described are indicated in parentheses):

Root depth observations are the most reliable indicator, with horizons incapable of supporting more than few fine or very roots (see Section 3.1.5c) considered root restricting.

Continuously cemented zones (see Section 3.1.6b) of any thickness are considered root restricting.

Zones >10 cm below the rooting zone are considered root restricting if they exhibit the following characteristics when water state is very moist or wet (Section 3.2.1): (1) structure is massive, platy, or has weak structure of any type for a vertical repeat distance of < 10 cm (Section 3.1.5a), and (2) rupture resistance is very firm (firm, if sandy) or extremely firm (Section 3.1.6b), or has a large penetration resistance (Section 3.1.6e).

When a root restricting layer is present, soils are classified according to the following depth classes:

Class	Depth
Very shallow	< 25 cm
Shallow	25 to 50 cm
Moderately deep	50 to 100 cm
Deep	100 to 150 cm
Very deep	> 150 cm

The type or kind of restrictive layers may be identified by horizon subscript codes (i.e. "d", "f", "m", "r", "v", and "x"), master horizon "R" (see Table 3-1) or miscellaneous layers designations as follows (Soil Survey Staff, 1993):

Code

dur — duripan
frag — fragipan
calc — petrocalcic
gyp — petrogypsic
lith — bedrock (lithic)
para — paralithic

fe — petroferric
iron — plinthite
ice — permafrost
natr - natric
sali — salic
sulf — sulfuric
plac — placic
orst — ortstein
abr — abrupt textural change
dens — dense material (glacial till)
sr — strongly contrasting textural stratification

Keys to Soil Taxonomy (Soil Survey Staff, 1992) should be referred to for more information on the characteristics of these restrictive layers. Most of the above listed restrictive layers significantly reduce downward movement of water in the soil. Some horizons, however, such as salic horizons (containing soluble salts) and some types of paralithic horizons (such as weathered diorite high in biotite) function as root-restricting layers, but may not significantly restrict downward movement of water. Other restrictive layers, such as abrupt textural changes and strongly contrasting textural stratification may cause temporary perching of water at the contact between the contrasting textures, but allow downward movement to continue when sufficient head builds for water to move across the contact.

3.1.6e Penetration Resistance (Compaction)

Penetration resistance is the capacity of the soil in its confined state to resist penetration by a rigid object. It is usually reported as megapascals (1 MPa = 10 bars or 9.9 atmospheres of pressure). Large penetration resistance is an indicator of compaction or other soil features that impede vertical flow of contaminants (see Section 3.1.6a and d). Truck or tractor-mounted cone penetrometers are commonly used in engineering investigations and are increasingly being used to characterize the subsurface at contaminated sites.

For soil descriptions, the pocket penetrometer, a hand-operated, calibrated-spring penetrometer, is a useful tool for helping identify root restricting layers. It is simple to operate, and can be pushed into the soil surface, core samples, or soil exposed in an open pit investigation.

Penetration resistance depends strongly on water state (Section 3.2.1), which should be specified. Orientation of the axis of insertion should also be specified. SCS defines penetration resistance classes based on the pressure required to push a pocket penetrometer with a diameter of 6.4 mm a distance of 6.4 mm into the soil in about 1 second, as follows:

Classes	Penetration Resistance (MPa)	
Small	<0.1	
Extremely low		<0.01
Very low		0.01-0.1
Intermediate	0.1-2	
Low		0.1-1
Moderate		1-2
Large	>2	
High		2-4
Very high		4-8
Extremely high		>8

Compacted near-surface zones resulting from equipment traffic or tillage will have large penetration resistance, and higher rupture resistance (Section 3.1.6b) and bulk densities (Section 3.1.6c) than undisturbed near-surface horizons.

Soil Survey Staff (1993) and Bradford (1986) provide further guidance on use and interpretation of pocket penetrometer readings. Fritton (1990) describes procedures for adjusting cone penetrometer data to allow comparison of measurements by cone penetrometers with different sizes or shapes.

EPA's ESES uses the following soil compaction classes:

High: Surface soils have been subject to high compaction and consequent effects on soil structure, such as by vehicular and foot traffic, or livestock.

Moderate: Surface soils have been less subjected to compaction, either through reduced applications and frequency of pressure on surface soils or because the soil structure is more resistant to compaction.

Low to Slight: Surface soils are only slightly affected by compaction, either because of resistance to compaction or because they are less subject to application of compacting stressor.

3.1.7 Soil Engineering Parameters and Properties

Texture, clay content (amount and types of clays), and strength behavior at different moisture contents are the key properties affecting soil engineering. A number of classification systems have been developed for the selection of soil materials and design of foundations and earthen structures. The ASTM (Unified) soil classification is the most widely used at contaminated sites and is the only one covered in this guide.

Only field procedures for preliminary estimation of soil engineering properties are covered here. Laboratory tests are required for accurate determination of soil engineering properties. Once the Unified soil class has been identified, other properties such as permeability and suitability for different types of engineering applications can be estimated using Figure 4-14 in SCS (1990).

3.1.7a Unified (ASTM) Texture

Form 3-2 can be used to record the results of 11 tests for field classification of soil texture for engineering uses. These tests are drawn from Brown et al. (1991), SCS (1990), and Soil Survey Staff (1993). Figure 3-8 summarizes how the results of these tests are used to estimate texture in the Unified soil classification system. Procedures for specific tests identified in Figure 3-8 are described below, along with some alternative versions described by Soil Survey Staff (1993):

Form 3-2. Unified (ASTM) Field Texture Determination Form (adapted from Brown et al., 1991)

Sample No. _____ Site: _____ Date: _____
Sampling Location: _____
_____

Person(s) Performing Test: _____
_____
_____

SANDS AND GRAVELS

Test 1 (coarse grained) _____
Test 2 (gravel)   _____
Test 3 (fine grained)   _____

COMMENTS:
_____
_____
_____
_____

SILTS AND CLAYS

Test 4 (plasticity)   _____
Test 5 (ribbon)   _____
Test 6 (liquid limit)   _____
Test 7 (clod strength) _____
Test 8 (dilatancy)   _____
Test 9 (toughness)   _____
Test 10 (stickiness)   _____
Test 11 (organic soils) _____

COMMENTS:
_____
_____
_____
_____

					Classification
Test 1 COARSE-GRAINED SOILS More than half of the material (by weight) is individual grains visible to the naked eye.	**Test 2a** GRAVELLY SOILS—More than half of coarse fraction is larger than 1/4".	**Test 2b** CLEAN GRAVELS Will not leave a stain on a wet palm	**Test 2c** Substantial amounts of all grain particle sizes		GW
				Predominantly one size or a range of sizes with intermediate sizes missing	GP
		Test 2b DIRTY GRAVELS Will leave a stain on a wet palm	**Test 4** Nonplastic fines		GM
				Plastic fines	GC
	Test 2a SANDY SOILS—More than half of coarse fraction is smaller than 1/4".	**Test 2b** CLEAN SANDS Will not leave a stain on a wet palm	**Test 2c** Wide range in grain size and substantial amounts of all grain particle sizes		SW
				Predominantly one size or a range of sizes with intermediate sizes missing	SP
		Test 2b DIRTY SANDS Will leave a stain on a wet palm	**Test 4** Nonplastic fines		SM
				Plastic fines	SC

Test 3 FINE-GRAINED SOILS More than half of the material (by weight) is individual grains not visible to the naked eye.	Test 5-RIBBON	Test 6-LIQUID LIMIT	Test 7-DRY CRUSHING STRENGTH	Test 8-DILATANCY REACTION	Test 9-TOUGHNESS	Test 10-STICKINESS	
	None	<50	None to Slight	Rapid	Low	None	ML
	Weak	<50	Medium to High	None to Very Slow	Medium to High	Medium	CL
	Strong	>50	Slight to Medium	Slow to None	Medium	Low	MH
	Very Strong	>50	High to Very High	None	High	Very High	CH

Test 11—HIGHLY ORGANIC SOILS	Readily identified by color, odor, spongy feel, and frequently by fibrous texture.				OL, OH, Pt

Figure 3-8. Summary of field tests for Unified (ASTM) soil textural classification (Source: SCS, 1990).

3-43

Test 1 - Coarse-Grained Soil Test

a.   Spread a sample of soil on a flat surface (clipboard) and examine the particles.

b.   Approximate the grain size by visual inspection.

c.   If **more** than 50 percent of the grains are easily distinguished by the unaided eye, the material is a coarse soil; if **less** than 50 percent of the grains are easily distinguished by the unaided eye, the material is a fine soil.

d.   If some of the particles could be aggregates of fine particles, saturate a small sample of the soil with water.

e.   Rub a large marble-sized (2 cm) sized sample between the thumb and forefinger. Sand grains (coarse material) will feel rough and gritty, whereas aggregates of fine materials will break down and feel silky.

Alternative Jar Method (Soil Survey Staff, 1993)

f.   Thoroughly shake a mixture of soil and water in a straight-sided jar or test tube, and allow the mixture to settle.

g.   Sand sizes will fall out first, in 20 to 30 seconds; successively finer particles will follow. Estimate the proportions of sand and fines from their relative volumes.

Test 2 - Sand/Gravel Tests

2.a Gravel/Sand

a.   Spread a representative sample of soil on a flat surface.

b.   If more than one-half of the visible grains are greater than 2 mm, the material is gravel; if they are not, it is a sand.

2.b Clean/Dirty Test

c.   Remove any coarse material greater than 2 mm in diameter.

d.   Saturate the remaining materials with water and work it with your hands.

e.   Hands will not be stained when fines are less than 5 percent (GW or GP; SW or SP); hands will be stained when there is more than 12 percent fines, and weak casts can be formed (GM or GC; SM or SC).

2.c Sorting Test

f.   Take a second sample of soil and place it on a flat surface. Spread it out and observe the grain size distribution.

g.   If coarse materials consist of evenly distributed particle sizes, the material is well graded (GW or SW); if they are chiefly of one size particle or large and small particles without intermediate sizes, the material is poorly graded (GP or SP).

Test 3 - Fine-Grained Test

a.   When Test 1 indicates fine-grained soils complete Tests 4 through 7 to determine if the material is clayey or silty.

b.   If Test 2b indicates dirty sands or gravels (> 12 percent fines), complete Test 4 to determine whether fines are plastic or nonplastic.

Test 4 - Plasticity Test

a.   Wet and mold a small 2 x 2 x 2 cm (3 teaspoons) soil sample so that it can be rolled into a thread without crumbling. The material will not stick to the hands if the correct amount of water is added.

b.   Roll the moist soil with the palm of the hand on any clean, smooth surface, such as a piece of paper or clipboard, to form a coarse thread and pull it apart.

c.   Observe difficulty of pulling thread apart: GC, SC, CH or CL = tough (hard to pull apart); GM, SM, or MH = medium tough; GM, SM, or ML = weak (easily pulled apart).

Alternative Plasticity Test (from Soil Survey Staff, 1993)

d.   Find the minimum thickness a 4-cm long roll must have to support its own weight: Nonplastic = > 6 mm; slightly plastic = 4-6 mm; moderately plastic = 2-4 mm; very plastic = < 2 mm.

Test 5 - Ribbon Test

a.   Take a quantity of soil measuring at least 2 x 2 x 2 cm (3 teaspoons). Square and wet it with water until it reaches a plastic state. This condition prevails when the soil contains just enough moisture so that it can be rolled into 3-mm diameter threads. These threads or ribbons are formed by squeezing and working the sample between the thumb and forefinger. Plastic limit is governed by clay content.

b.   Observe properties of ribbon: ML = weak (breaks easily); MH = hard (does not break easily); CL = flexible with medium strength; CH = strong and flexible.

Test 6 - Liquid Limit Test (from SCS, 1990)

a.   Take a pat of moist soil with a volume of about 8 cc (1.2 cu.in.) and add enough water to make the soil soft but not sticky.

b.   Rapidly add enough water to cover the outer surface, and break the pat open immediately.

c.   A positive reaction has occurred when the water has penetrated through the surface layer: LL = low, if water has penetrated (ML, CL); LL = high, if the water has not penetrated (MH, CH). [Note: direct sunlight facilitates observation of this phenomena].

Test 7 - Dry Crushing/Clod Strength Test

a.   Obtain a dry block of soil at least 2 cm (3/4 in.) in its smallest dimension.

b.   Crush the clod between fingers and observe the effort required: ML = easily crushed; CL or MH = medium-hard to break; CH = almost impossible to break.

Test 8 - Dilatancy Test (from SCS, 1990)

a.   Mold material into a ball about 15 mm (1/2 in.) in diameter. Add water, if needed, until it has a soft but not sticky consistency.

b.   Smooth the soil in the palm of one hand with the blade of a knife or spatula.

c.   Shake horizontally, striking the side of the hand against the other several times. Note the appearance of water on the surface. Squeeze the sample and note the disappearance of water.

d.   Describe the reaction: None = no visible change (CH); slow = water appears slowly on the surface during the shaking and does not disappear or disappears slowly when squeezed (CL, MH); rapid = water appears quickly on the surface during shaking and disappears quickly when squeezed (ML).

Test 9 - Toughness Test (from SCS, 1990)

a.   Take the specimen for the dilatancy test and shape it into an elongated pat and roll it on a hard surface or between hands into a thread about 3 mm (1/8 in.) in diameter. (If it is too wet to roll, spread it out and let it dry.)

b. Fold the thread and reroll repeatedly until the thread crumbles at a diameter of 3 mm (1/8 in.). The soil has reached its plastic limit.

c. Note the pressure required to roll the thread and the strength of the thread: circumferential breaks = CH or CL material; longitudinal cracks and diagonal breaks = MH.

d. After the thread crumbles, lump the pieces together and knead until the lump crumbles.

e. Note the toughness of the material during kneading. Describe toughness: low = only slight pressure required to roll the thread near the plastic limit, the thread and lump are weak and soft (ML); medium = medium pressure is required to roll the thread near the plastic limit, the thread and lump have medium stiffness (CL, MH); high = considerable pressure required to roll the thread near the plastic limit, the thread and lump are very stiff (CL, CH); nonplastic = thread cannot be rolled (CL, CH).

Test 10 - Stickiness Test

a. Saturate a sample of the soil and let it dry on the hands.

b. Observe ease with which soil is rubbed off: ML is brushed off with little effort; CL or MH require moderate effort to brush off; CH requires rewetting to completely remove.

Alternative Stickiness Test (from Soil Survey Staff, 1993)

c. Saturate a sample of the soil and press between thumb and forefinger.

d. Observe the adhesion to thumb and forefinger when they are pulled apart: nonsticky = practically no adhesion (ML); slightly sticky = sticks to thumb or forefinger (MH); sticky =

adheres to both, stretches slightly before breaking (CL); very sticky = adheres to both, stretches decidedly (CH).

Test 11 - Organic Soils Test

a.   Smell soils suspected of having a high organic matter content.

b.   A distinctive, pungent musty odor is indicative of organic soils.

c.   Feel texture of soil: PT = spongy or fibrous texture; OL or OH = nonfibrous.

d.   If nonfibrous, do plasticity test (Test 6): OL = low plasticity; OH = high plasticity.

3.1.7b Atterberg Limits

Atterberg limits define various states of fine-grained soil material ranging from dry to liquid. The **shrinkage limit** (SL) is the water content at which a further reduction in water does not cause a decrease in the volume of the soil mass. The **plastic limit** (PL) is the water content at which soil changes from a semi-solid to a plastic state. At the plastic limit, a fine-grained soil will just begin to crumble when rolled into a thread approximately 3 mm (1/8 in.) in diameter. The **liquid limit** is the water content at which soil consistency changes from a plastic to a liquid state. This is the water content at which a pat of soil, cut by a groove 2 mm wide will flow together for a distance of 13 mm (1/2 in.) under the impact of 25 blows in a standard liquid limit apparatus.

Accurate determination of Atterberg limits require collection of samples for laboratory analysis.

3.1.7c Shear Strength

Shear strength can be estimated approximately in the field by the ease with which a sample can be penetrated by the thumb, as described in Table 3-6.

Table 3-6. Field Estimation of Soil Shear Strength

Consistency	Identification Procedure	Shear Strength (Tons/ft^2 or kg/cm^2)
Soft	Easily penetrated several inches by thumb	<0.25
Firm	Penetrated several inches by thumb with moderate effort	0.25 to 0.50
Stiff	Readily indented by thumb, but penetrated only with great effort	0.50 to 1.00
Very Stiff	Readily indented with thumbnail	1.00 to 2.00
Hard	Indented with difficulty by thumbnail	>2.00

Source: SCS (1990)

3.1.7d Shrink-Swell

Certain clays shrink when pore water is lost during drying and subsequently swell during wetting. High shrink-swell clays are of concern at contaminated sites because deeply penetrating cracks create pathways for contaminant movement during the early stages of the wetting phase. Where sodium rich clays with high dispersivity exist in the subsurface, a form of erosion called piping is also a concern.

3-50

SCS defines shrink-swell potential classes based on linear extensibility (LE (%) = 100 x moist length - dry length/dry length) or the coefficient of linear extensibility (COLE = LE/100), as follows:

Class	LE (%)	COLE	Dbd/Mdb
Low	<3	<0.03	<1.1
Medium	3-6	0.03-0.06	1.1-1.2
High	6-9	0.06-0.09	1.2-1.3
Very High	>9	>0.09	>1.3

LE can be measured by collecting a core from a wet or moist soil, carefully measuring its wet length, and setting it upright in a dry place. If the sample shrinks in a symmetrical shape without excessive cracking or crumbling, its length should be measured and LE calculated. If the core crumbles or cracks, the coated clod method for bulk density measurement (see Section 3.1.6c) and the ratio of dry bulk density (Dbd) to moist bulk density (Mbd) should be used to place samples in shrink-swell classes (see table above). SCS (1971, Section I4.3) and SCS (1983, 1992) provide additional information on these procedures. SCS (1971) also describes a simple test for rough determination of maximum potential shrinkage and density of disturbed soils.

3.1.7e Corrosivity

SCS soil series interpretation sheets give separate corrosion potential ratings for uncoated steel and for concrete. If classification of a soil is uncertain, corrosivity can be estimated using the following field and laboratory-measured parameters:

Uncoated steel — Texture (Section 3.1.2), water table (Section 3.2.2), drainage class (Section 3.2.4), total acidity or extractable acidity as a rough equivalent to total acidity (Section 3.3.3), resistivity at field capacity, and conductivity of saturated extract (Section 3.3.6).

Concrete — Texture (Section 3.1.2), reaction (Section 3.3.4), and laboratory analysis of sodium and/or magnesium sulfate (ppm) and sodium chloride (ppm).

Table 3-7 shows criteria for soils with **high, moderate,** and **low** corrosion potential for uncoated steel. Table 603-8 in SCS (1983) or Table 618-2 in SCS (1992) provide criteria to estimate corrosion potential for concrete.

3.1.8 Soil Temperature/Temperature Regime

Soil temperature affects evaporation rates of water and volatile contaminants and influences the amount of microbiological activity in the soil. In soil classification, mean annual or summer soil temperature and the relationship between mean summer and mean winter soil temperature is used for placement in soil temperature regimes.

Ground-water temperature gives a close estimate of mean annual soil temperature if monitoring wells are available with water at a depth of 10 to 20 meters. Alternatively, the average of four temperature measurements taken at a depth of about 50 cm equally spaced throughout the year gives a good estimate of mean annual soil temperature. For mean summer temperature, readings may be taken on June 15, July 15, and August 15, and for mean winter temperature, on December 15, January 15, and February 15 .

SCS (1971, Section I4.4) and Smith et al. (1960) describe more detailed procedures for soil temperature regime measurement, including special procedures where ground water or bedrock is shallow. Taylor and Jackson (1986) discuss types of thermometers and methods for measuring soil temperature for other applications.

EPA's ESES defines the following soil temperature classes based on point thermal measurements taken on the surface or with depth of soil at a point in time, or over a period of time:

High: $>38^{\circ}C$ $(100.4^{\circ}F)$

Medium: $8-38^{\circ}C$ (46.4 to $100.4^{\circ}F$)

Low: $<8^{\circ}C$ $(46.4^{\circ}F)$

Table 3-7. Guide for Estimating Risk of Soil Corrosion Potential of Uncoated Steel

Class	Drainage Class and Texture	Total Acidity* (meq/100g)	Resist-ivity* (Ohm/cm)	Conduct-ivity* (mmhos/cm)
Low	Excessively drained coarse-textured or well-drained coarse- to medium-textured soils; or, moderately well drained, coarse-textured soils; or, somewhat poorly drained coarse-textured soils	<8	≥5,000	<0.3
Moderate	Well drained, moderately fine-textured soils; or moderately well drained, coarse- and medium-textured soils; or somewhat poorly drained, moderately coarse-textured soils; or, very poorly drained soils with stable high water table	8-12	2,000 to 5,000	0.3 to 0.8
High	Well drained, fine-textured or stratified soils; or, moderately well drained fine-and moderately fine-textured or stratified soils; or, somewhat poorly drained, medium- to fine-textured or stratified soils; or poorly drained soil with fluctuating water table	>12	<2,000	≥0.8

* See Table 603-7 in SCS (1983) or Table 618-1 in SCS (1992), for further guidance on how measurements should be made and interpreted.

3.2 Soil Hydrologic Parameters

3.2.1 Moisture Conditions (Soil Water State)

The terms used to describe soil properties may vary depending on the soil moisture content of the sample being observed. Soil water state is a more precise term for moisture conditions (Soil Science Society of America, 1987). Soil water state should be noted for any observations of properties that may vary with moisture content, such as color and consistency.

Table 3-8 describes three main soil water state classes (dry, moist, and wet) and eight subclasses within these categories, as defined by SCS. The water state classification of a particular soil sample will depend on (1) texture (whether it is coarse or not), (2) how strongly water is held by the soil (suction), and (3) the amount of gravimetric water held by the soil in relation to the amount held at specified reference suction pressures (i.e., 1,500 kPa for dry soils).

Accurate placement of moist and dry classes requires field instrumentation. Referring to Table 3-8, the following can be used as guidelines for field estimation of soil water state:

Very dry — Very little visual or tactile change between field observation and after air-dried samples.

Moist — Visual or tactile change between field observation and after air drying.

Wet — Water films evident, but no free water (WN); free water present (WA).

Between moist/wet and moderately dry/very dry separations, Soil Survey Staff (1991) describes four field water state tests: (1) color value test, (2) ball test, (3) rod test, and (4) ribbon test. These tests require calibration with the results of tests on similar soils that have been conducted at different known points on the moisture retention curve.

Table 3-8. Water State Classes

Class	Criteria Suction (kPa)*	Retention (% gravimetric moisture)
Dry (D)	>1500	
Very dry (DV)		<.35 x % water at 1500 kPa
Moderately dry (DM)		.35 - 0.8 x % water at 1500 kPa
Slightly dry (DS)		0.8 to 1.0 x % water at 1500 kPa
Moist (M)	>1-1500	
Slightly moist (MS)		% water at 1500 kPa to MWR
Moderately moist (MM)		MWR** to UWR**
Very moist (MV)		UWR** to 1 or 0.5 kPa
Wet (W)	1	
Nonsatiated (WN)		No free water
Satiated (WA)		Free water present

Source: Adapted from Soil Survey Staff (1993).

*Coarse soil material is considered wet at 0.5 kPa suction and moist from 0.5 to 1500 kPa if it meets the following criteria: (1) sand or sandy-skeletal family particle size family, (2) coarser than loamy fine sand, (3) <2 percent organic carbon, (4) <5 percent water at 1500 kPa, and (5) computed total porosity of the <2 mm fabric exceeding 35 percent.

**UWR = upper water retention = % water at 5 Kpa (coarse soil) = % water at 10 KPa (other soil). MWR = midpoint water retention = halfway between UWR and % water at 1500 Kpa.

The following rules of thumb can be used to estimate actual percentages of water at different points on the moisture characteristic curve:

Water (%) at 1,500 Kpa = 0.4 x estimated clay content

Water (%) of air dry soil = 0.1 x estimated clay content

Methods for accurately measuring moisture content and matric potential in the field require relatively sophisticated equipment and are not discussed further here (see Appendix C for information on field methods). The same is true for accurate measurement of water retention at different matric potentials (moisture characteristic curves).

Moisture content and moisture characteristic curves are most commonly measured in the laboratory from samples collected in the field. Soil moisture tins with a capacity for about 250 grams of soil, filled to the brim and sealed with tape or plastic bags with any extra air removed before sealing are best for determining field moisture content at the time of sampling. Samples as small as 50 grams placed in air tight containers may be acceptable (Brown et al., 1983). Undisturbed core samples are best for laboratory measurement of moisture characteristic curves.

SCS (1971, Section I4.1) describe procedures for determining water content that can be done in a field laboratory with the following equipment: balance accurate to 0.1 g; oven with thermometer or thermostat, or stove, hot plate, or infrared lamp and thermometer with scale up to 150 degrees C.

3.2.2 Water Table (Internal Free Water Occurrence)

SCS does not define a class for saturation (i.e., zero air-filled porosity) because the term implies that all of the pore space is filled with water, a situation which cannot be readily evaluated in the field. Free water develops positive pressure when its depth is below the top of a wet satiated zone (the top of water in an unlined borehole after equilibrium has been reached).

SCS classifies free water occurrence (perched or regional water table) into classes depending on (1) thickness (if perched), (2) depth, and (3) duration as follows:

Classes	Criteria
Thickness if perched	
Extremely Thin (TE)	< 10 cm
Very Thin (TV)	10 to 30 cm
Thin (T)	30 cm to 1 m
Thick (TK)	> 1 m
Depth	
Very Deep (DV)	> 1.5 m
Deep (D)	1.5 to 1 m
Moderately Deep (DM)	1 to 0.5 m
Shallow (S)	0.5 m to 25 cm
Very Shallow (SV)	< 25 cm
Cumulative Annual Duration	
Absent (A)	Not observed
Very Transitory (TV)	Present < 1 month
Transitory (T)	Present 1 to 3 months
Common (C)	Present 3 to 6 months
Persistent (PS)	Present 6 to 12 months
Permanent (P)	Present continuously

3.2.3 Available Water Capacity

Available water capacity (AWC) is the amount of plant-available soil moisture, usually expressed as inches of water per inch of soil depth. It is commonly defined as the amount of water held between field capacity (FC) and the wilting point (percent water at 1500 kPa):

AWC (in/in) = (% water in < 2mm fraction at FC - % water at 1500 kPa)/100 x moist bulk density x (1 - volume > 2 mm)/100

$$\text{AWC (in/in/horizon)} = \text{AWC (in/in)} \times \text{horizon thickness}$$

AWC is important in developing water budgets and designing drainage systems.

Accurate determination of AWC requires sampling of soil for moisture content when it is at field capacity. This requires sampling just after the soil has drained after a period of rain and humid weather, after a spring thaw, or after heavy irrigation. SCS (1971, Section I4.1) describes several other procedures for bringing a plot to field capacity by wetting, and several methods for wetting clods or cores to approximate field capacity. Field capacity can be approximated in the laboratory by measuring moisture percentage at 33 kPa for clayey and loamy soil materials and 10 kPa for sandy materials.

Classes of available water capacity are based on the sum of AWC in inches to a specified depth, which should be based on the typical rooting depth of the plants or crops of interest. SCS (1992) presents the following classes, which would be appropriate for a crop such as corn:

Classes	AWC (in/60 in)
Very low	0 to 3
Low	3 to 6
Moderate	6 to 9
High	9 to 12
Very high	12 to 42

SCS (1983, 1992) provide the following additional guidance for estimating total AWC in a soil profile: (1) restrictive layers that exclude root (Section 3.1.6d) and horizons below have AWC = 0; (2) non-porous rock fragments reduce AWC in proportion to the volume they occupy, but some estimation of plant-available capillary water in porous rock fragments may be required; (3) a rough guideline for reducing AWC in saline soils is to reduce it by about 25 percent for each 4 dS/m of electrical conductivity of the saturated extract (Section 3.3.6); (4) AWC in soil high in gibbsite or kaolinite (e.g. Oxisols and Ultisols) may be about 20 percent lower than those with 2:1 lattice clays; (5) soils high in organic matter have higher AWC than soils low in organic matter, other properties being the same.

3.2.4 Saturated Hydraulic Conductivity and Soil Drainage Class

The terms permeability and saturated hydraulic conductivity are often used interchangeably to refer to the ease with which water moves through the soil under saturated conditions. Permeability rates are typically reported in units of in./hr based on percolation tests; saturated hydraulic conductivity may be reported in units of μ/s, m/s, cm/day, in./hr, or cm/hr. Accurate field measurement of both saturated (K_{sat}) and unsaturated hydraulic conductivity requires relatively complex instruments and procedures that are not covered here (see Appendix C).

SCS currently defines six classes for describing soils based on saturated hydraulic conductivity:

Class	Saturated Hydraulic Conductivity (μ/s)	(in./hr)
Very Low (VL)	<0.01	<0.001
Low (L)	0.01-0.1	0.001-0.01
Moderately Low (ML)	0.1-1	0.01-0.14
Moderately High (MH)	1-10	0.14-1.4
High (H)	10-100	1.4-14.2
Very High (VH)	>100	>14.2

Class placement is based on geometric mean of multiple measurements.

Various methods have been developed for obtaining rough estimates of K_{sat} based on various soil properties. Table 3-9 provides a basis for estimating Ksat based on field observations of structure, texture, and pores. See Section 3.1.2 for identification of textural classes, Section 3.1.4 for pore definitions, and Section 3.1.5a for definitions of structural units.

Table 3-9. Guide for Estimating the Class of Saturated Vertical Hydraulic Conductivity from Soil Properties

Class Name	Rate*	Soil Properties**
Very High	>100	—Fragmental, cindery, pumiceous —Sandy and sandy-skeletal with coarse sand or sand texture, and loose consistence —More than 0.5 percent medium or coarser vertical pores with high continuity
High	100-10	—Other sandy, sandy-skeletal, coarse-loamy or Andic soil materials that are very friable, friable, soft, or loose —When very moist or wet has moderate or strong granular structure; or strong blocky structure of any size or prismatic structure finer than very coarse, and many surface features except stress surfaces or slickensides on vertical surfaces of structural units —0.5 to 0.2 percent medium or coarser vertical pores with high continuity
Moderate	10-1	—Other sandy and Andic soil materials in other consistence classes except extremely firm or cemented —18 to 35 percent clay with moderate structure, except platy, or with strong very coarse prismatic structure; and with common surface features except stress surfaces or slickensides on vertical surfaces of structural units —0.1 to 0.2 percent medium or coarser vertical pores with high continuity

(continued)

Table 3-9 (Continued)

Name	Class Rate*	Soil Properties**
Moderately Low	1-0.1	—Other sandy classes that are extremely firm or cemented —18 to 35 percent clay with other structures and surface conditions except pressure or stress surfaces —≥ 35 percent clay and moderate structure except if platy or very coarse prismatic; and with common vertical surface features except stress surfaces or slickensides —Medium or coarser vertical pores with high continuity, but < 0.1 percent
Low	0.1-0.01	—Continuous moderate or weak cementation —Greater than 35 percent clay and meets one of the following: weak structure; weak structure with few or no vertical surface features; platy structure; common or many stress surfaces or slickensides
Very Low	< 0.01	—Continuously indurated or strongly cemented and with less than common roots —Greater than 35 percent clay and massive or exhibits horizontal depositional strata and less than common roots

Source: SCS (1992).

* μm/second.

** A given soil profile whould have most, but not necessarily all of the soil properties associated with a particular class.

O'Neal (1952) describes a procedure for somewhat more precise estimation of soil permeability classes (which are defined slightly differently from those listed above) based on (1) structure, (2) shape and overlap of aggregates, (3) visible pores, and (4) texture. Soil Survey Staff (1993, Chapter 3) provides figures for Ksat class placement based on soil bulk density and texture.

Soil drainage class refers to the frequency and duration of wet periods for the water regime associated with undisturbed soil conditions. Table 3-10 summarizes SCS criteria for seven soil drainage classes.

3.2.5 Infiltration

The amount of precipitation reaching the ground surface that enters the soil is determined by the **infiltration capacity.** In a dry soil, infiltration is usually rapid, unless there is an impervious crust at the surface. As time passe, infiltration slows until the **ponded infiltration** rate is attained, which is determined by the saturated hydraulic conductivity of the soil. Infiltration rate is usually measured in rates of in./hr or cm/hr. Extrastructural cracks (Section 3.1.5b) may greatly increase infiltration rates compared to soils with similar texture which do not have cracks.

SCS's permeability classification system (SCS, 1992) can be used to describe infiltration classes:

| | **Permeability** |
Class (in/hr)	(cm/hr)
Very Slow	<0.06<0.15
Very Extremely Slow	<0.01
Extremely Slow	0.01-0.06
Slow 0.06-0.2	0.15-0.5
Moderately Slow	0.2-0.60.5-1.5
Moderate	0.6-2.01.5-5.0
Moderately Rapid	2.0-6.05.0-15.2
Rapid 6.0-20	15.2-50.8
Very Rapid	>20>50.8

Table 3-10. Criteria for SCS Soil Drainage Classes

Excessively drained (E). Water is removed very rapidly. Internal free water occurrence commonly is very deep; annual duration is not specified. The soils are commonly very coarse textured, rocky, or shallow. Some are steep. All are free of mottling related to wetness.

Somewhat excessively drained (SE). Water is removed from the soil rapidly. Internal free water occurrence commonly is very deep; annual duration is not specified. The soils are usually sandy and rapidly pervious. Some are shallow. A portion of the soils are so steep that a considerable part of the precipitation received is lost as runoff. All are free of the mottling related to wetness.

Well drained (W). Water is removed from the soil readily but not rapidly. Internal free water occurrence commonly is deep or very deep; annual duration is not specified. Water is available to plants throughout most of the growing season in humid regions. Wetness does not inhibit growth of roots for significant periods during most growing seasons. Well drained soils are commonly medium textured. They are mainly free of the mottling related to wetness.

Moderately well drained (MW). Water is removed from the soil somewhat slowly during some periods of the year. Internal free water occurrence commonly is moderately deep (0.5 to 1 m) and transitory through permanent (Section 3.2.2). The soils are wet for only a short time during the growing season, but long enough that most mesophytic crops are affected. They commonly have a slowly pervious layer within the upper 1 m, and periodically receive high rainfall or both. *Color*: Low chroma mottling between 30 to 100 cm and lack yellowish soil matrix hues or neutral colors within 150 cm.

Somewhat poorly drained (SP). Water is removed slowly enough that the soil is wet at shallow depth for significant periods during the growing season. Internal free water occurrence commonly is shallow (25 to 50 cm) and transitory or common. Wetness markedly restricts the growth of mesophytic crops unless artificial drainage is provided. The soils commonly have one or more of the following characteristics: contain a slowly pervious layer, have a high water table, receive additional water from seepage, or occur under nearly continuous rainfall. *Color*: Low chroma mottling, yellowish soil matrix hues or low chroma or neutral colors that make up <60% of the soil matrix color between the A or Ap horizon and a depth of 75 cm or to a fragipan.   (continued)

Table 3-10 (Continued)

Poorly drained (P). Water is removed so slowly that the soil is wet at shallow depths periodically during the growing season, or remains wet for long periods. Internal free water occurrence is shallow (25 to 50 cm) or very shallow (<25 cm) and common or persistent. Free water is commonly at or near the surface for long enough during the growing season that most mesophytic crops cannot be grown unless the soil is artificially drained. The soil, however, is not continuously wet directly below plow-depth. Free water at shallow depth is usually present. This water table is commonly the result of a shallow, slowly pervious layer within the soil, of seepage, of nearly continuous rainfall, or of a combination of these. *Color*: low chroma mottling, yellowish or bluish soil matrix hues, or low chroma or neutral colors that make up 60% or more of the soil matrix between the A or Ap horizona and a depth of 75 cm or to a fragipan.

Very poorly drained (VP). Water is removed from the soil so slowly that free water remains at or very near the ground surface during much of the growing season. Internal free water occurrence is very shallow (<25 cm) and persistent or permanent (Section 3.2.2). Unless the soil is artificially drained, most mesophytic crops cannot be grown. The soils are commonly level or depressed and frequently ponded. If rainfall is high or nearly continuous, slope gradients can be moderate or high. *Colors*: from soil surface down, low chroma mottling, yellowish or bluish soil matrix hues, or low chorma or neutral colors throughout the soil.

Source: Soil Survey Staff (1993); color criteria from Langan and Lammers (1991).

3.3 Soil Chemistry and Biology

Most procedures for characterization of soil chemistry require collection of samples for laboratory analysis. This section focuses on pH and mineralogical parameters that can be tested in the field using relatively simple equipment and procedures. Most of these tests are drawn from SCS (1971).

3.3.1 Organic Matter

Organic matter (more precisely measured in the laboratory as organic carbon) affects contaminant mobility primarily by its high sorptive capacity. Accurate determination of total organic carbon (TOC) requires collection of samples for laboratory analysis.

In the field, black or dark colors generally indicate high organic matter content in near-surface horizons. In most mineral soils, organic matter is moderately low to high in the A horizon, and low in the subsoil. Severely eroded soils, where the topsoil has been completely removed; alluvial soils, where flooding deposits topsoil; and buried soils may not follow this pattern. SCS defines the following classes for organic matter:

Class	% Organic Matter
Very low (VL)	<0.5
Low (LL)	0.5-1.0
Moderately low (ML)	1.0-2.0
Medium (MM)	2.0-4.0
High (HH)	>4.0

Note that organic matter content is generally 1.7 to 2.0 times the TOC. An ignition test can be used in the field to approximate organic matter content and check its contribution to soil color. Equipment for this test includes (1) thermometer and heat lamp, (2) portable gas soldering torch, (3) a porcelain crucible or small tin can (not aluminum), (4) wire brackets or tongs to hold the container, and (5) a balance accurate to 0.1 grams.

Ignition Test Procedure (adapted from SCS, 1971)

1. Dry about 30 grams of soil to 110 degrees C under heat lamp.

2. Weigh as accurately as possible about 10 grams of dried soil and place in crucible or tin, supporting it with tongs or wire bracket.

3. Apply the flame of the torch to the bottom and lower walls of the outside of the container. Porcelain and metal glow red at 500°C. At this temperature, organic matter is completely burned and the water of hydration is removed from the common oxide and clay minerals.

4. Observe changes in color in the specimen. Apply heat more than once until there is no more change apparent in the specimen. Do not apply the flame directly to the sample if burning or oxidation are the purpose of the test, because unpredictable reducing conditions exist in parts of the torch flame. If organic matter is the only material giving color to the soil, it burns away leaving a whitish residue. See Section 3.1.3 for interpretation of other color changes.

5. When no more change is apparent, cool and weigh sample again. The loss in weight divided by the original weight times 100 equals the organic matter in sandy soils and materials high in organic matter. If much clay is present, the loss also includes water of hydration in the minerals.

Organic matter also reacts with hydrogen peroxide. In contrast to manganese oxides (see Section 3.3.8), the reaction starts slowly, builds up, and continues. Organic matter reactions decrease with depth, whereas manganese oxide reactions remain constant.

Test No. 11 in Section 3.1.7a describes procedures for classifying organic soils in the Unified soil classification system. SCS uses a number of tests for classification of organic soils (Histosols) including sodium pyrophosphate color, fiber percentages, and pH in 0.01 M calcium chloride. Since organic soils do not commonly occur at hazardous waste sites, these tests are not described here. If organic soils are present at a site and more detailed characterization is required, refer to Lynn et al. (1974) and Appendix III in Soil Survey Staff (1975).

3.3.2 Odor

High organic matter content in soil is associated with a distinctive, pungent musty odor (see Test 11 in Section 3.1.7a). Organic rich topsoil in mineral soils can also be distinguished from subsoil that is low in organic matter by this odor.

Volatile organic contaminants can impart distinctive odors to soil. Gasoline has an odor familiar to most people; aging gives petroleum a musty odor. Some other contaminants that may give soil a noticeable odor include halogens, ammonia, turpentine, phenols and cresol, picrates, various hydrocarbons, and unsaturated organic pesticides.

Caution should be exercised in observing soil odors. They should not be vigorously inhaled from any soil, since even natural soils may contain potentially harmful microorganisms. Soils where noticeable artificial odors are present should be checked for volatile concentrations using a detection instrument (HNu, organic vapor analyzer, etc.). The health and safety officer should be consulted to determine whether special protective equipment, such as a respirator, should be used by individuals who are working close to the soil surface taking samples.

EPA's ESES uses the following odor classes:

High: A distinct odor, from naturally occurring soil organic materials with a distinctive pungent, musty odor, or sharp distinct odor from chemical contaminants.

Moderate to slight: A less distinct to faint odor, from naturally occurring soil organic materials, or from various odor producing chemical contaminants.

None: No detectable odor by olfactory means.

3.3.3 Cation Exchange Capacity (CEC)

Cation exchange capacity, measured in milliequivalents per 100 grams (meq/100 g) or centimoles per kilogram (cmol/kg), is a measure of the soil's ability to absorb (and release) cations. It is an especially important parameter at sites contaminated by heavy metals, because heavy metals will often replace exchangeable ions such as sodium, potassium, calcium, and magnesium that exist in natural soil. Measurement of CEC requires collection of samples for laboratory analysis. SCS (1971, Section I9.7 and I9.8) describes relatively simple chemical tests that can be carried out in the field to estimate exchangeable calcium and exchangeable sodium.

The measurement of **extractable acidity**, also called exchangeable acidity or extractable hydrogen because it measures exchangeable ions that contribute to soil acidity, is required for evaluation of soil corrosivity (see Section 3.1.7e). Section 6H in SCS (1984) describes specific procedures for determining extractable acidity in the laboratory.

EPA's ESES uses the following CEC classes (expressed as meq/100 g soil): **High** (>20), **medium** (12-20), **low** (<12).

3.3.4 Reaction (pH)

A variety of methods are available for field measurement of pH (colorimetric, paper test strips, pH meter). Specific procedures and instructions accompanying equipment for the method used should be followed. For RCRA sites, EPA Method 9045A Revision 1, November 1990, should be used (U.S. EPA, 1986).

Soil Survey Staff (1993) defines 13 pH classes for soil as follows:

Class	pH
Ultra acid (UA)	<3.5
Extremely acid (EA)	3.5 - 4.4
Very strongly acid (VS)	4.5 - 5.0
Strongly acid (SA)	5.1 - 5.5
Moderately acid (MA)	5.6 - 6.0
Slightly acid (SA)	6.1 - 6.5
Neutral (NA)	6.6 - 7.3
Slightly alkaline (MA)	7.4 - 7.8
Moderately alkaline (MO)	7.9 - 8.4
Strongly alkaline (SA)	8.5 - 9.0
Very strongly alkaline (VA)	>9.0

3.3.5 Redox Potential (Eh)

Redox, or oxidation-reduction potential (Eh), is measured in volts or millivolts (mV) as the potential difference in a solution between a working electrode, and the standard hydrogen electrode. Whether soil conditions are oxidizing (aerobic) or reducing (anaerobic) will strongly affect the types of microbiological activity and contaminant transformation and degradation processes that may occur. The mobility of many heavy metals varies with oxidation state. In unsaturated soil, aerobic conditions prevail, and measurement of Eh requires collection of soil water samples using a suction lysimeter. The Eh of saturated soil should be measured using ground-water samples from properly purged monitoring wells.

EPA's ESES uses the following redox potential classes: **Highly oxidized** (> +400 mV), **intermediate** (+400 to -100 mV), and **highly reduced** (< -100 mV).

Appendix D provides more detailed information on description and interpretation of redoximorphic soil features.

3.3.6 Electrical Conductivity (Salinity)

In arid and semi-arid areas, soluble salts may accumulate in the soil causing saline conditions. The electrical conductivity of a saturation extract is the standard measure of salinity. Electrical conductivity is commonly reported in units of decisiemens/meter or millimhos/centimeter (1 dS/m = 1 mmho/cm). Electrical conductivity measurement or estimation is required for evaluation of soil corrosivity (see Section 3.1.7e).

Salinity classes, based on electrical conductivity of a saturation extract, are defined in EPA's ESES as follows:

Class	Electrical Conductivity (dS/m or mmho/cm)
Nonsaline	0-2
Slightly saline	2-4
Moderately saline	4-8
Very saline	8-16
Extremely saline	> 16

Note that Soil Survey Staff (1993) uses slightly different terms for these classes.

SCS (1971, Section I9.6) describes a relatively simple procedure for water extraction that can be used in the field or a field laboratory for measuring approximate soluble salt percentage. Accurate determination requires laboratory preparation of soil samples. SCS (1984, Section 8A) describes procedures for preparing a saturated paste and obtaining a saturation extract for electrical conductivity measurement. Measurement of the resistivity of the saturated paste (SCS, 1984, Section 8E) is required for evaluating soil corrosion potential for uncoated steel.

Richards (1954) and Richards et al. (1956) describe procedures for testing saline soils in more detail and provide charts and graphs for estimating total salt from electrical conductivity measurements. Rhoades and Oster (1986) describe more complex instrumentation for collecting soil water using in situ samplers and measuring soluble salts with in situ or remote monitors.

The electrical conductivity, also termed specific conductance, of ground-water samples can be readily measured in the field using a conductivity meter. Electrical conductivity of ground water is an indicator of the total dissolved solids content and should not be confused with the soil salinity test described above.

3.3.7 Clay Minerals

The following procedures can be used to identify the dominant mineralogy of the clay size-fraction:

Clay Mineral Test

1. Prepare a saturated solution of malachite green in nitrobenzene to use as an indicator solution (follow prescribed safety procedures when handling nitrobenzene).

2. Add several drops of the indicator to a small sample (1 g) of soil and observe color of wetted soil as follows: blue or green-blue = kaolinite; yellow-red = montmorillonite; purple-red = illite.

Mica Shine Test

Rub a small clod of air-dried soil with a knife blade. A shiny surface indicates a micaceous soil with high plasticity.

The degree of shrink-swell in a soil (Section 3.1.7d) also serves as an indicator of clay mineralogy: high = montmorillonitic; medium = illitic; low = kaolinitic.

EPA's ESES defines the following clay mineral abundance classes (see Section 3.1.2 for estimation of texture): **Abundant** (>27 percent), **moderate to slight** (1 to 27 percent), **none to negligible** (<1 percent).

3.3.8 Other Minerals

Concentrations of minerals may form in soil horizons as a result of dissolution and precipitation processes. These concentrations are described by SCS according to (1) type, (2) amount or quantity, (3) size, (4) shape, and (5) composition. The following are abbreviations of descriptors and criteria for describing concentrations in soil:

Types

m — masses (soft, no clearly defined boundaries)
n — nodules and concretions (hard, clearly defined boundaries)
c — crystals (single or complex clusters)
srf — soft rock fragments (weakly cemented or noncemented)

Amount/Quantity	Size
f — few (<2)	fine (<2 mm)
c — common (2-20%)	medium (2-5 mm)
m — many (>20%)	coarse (5-20 mm)
	very coarse (20-76 mm)
	extremely coarse (>76mm)

Shape	Composition
rnd — rounded	calc — calcareous
cyl — cylindrical	arg — argillaceous
pl — platelike	gyp — gypsiferous
ir — irregular	sil — siliceous
	fe — ferruginous
	mn - manganiferous
	sal - saline

The following features and tests can be used to identify major non-clay minerals in the field.

Carbonates (calcareous)

The presence of free calcium carbonate in soil can be readily determined based on effervescence in dilute hydrochloric acid. The test procedure is as follows:

1. Place 1 g of soil material (about the size of a marble) in the well of a porcelain spot plate. Thoroughly moisten the soil with a few drops of deionized water; stir with a clean glass rod to remove entrapped air.

2. Add three drops of dilute 10 percent (4 N) cold HCl from a plastic squeeze bottle and immediately observe for effervescence of the treated sample under a hand lens if possible.

3. If effervescence is observed, record intensity as follows:

    vse — very slightly effervescent (few bubbles seen)
    sle — slightly effervescent (bubbles readily seen)
    ste — strongly effervescent (bubbles form low foam)
    ve — violently effervescent (thick foam forms quickly)

4. Repeat procedure on a second 1-g sample.

Dolomite (calcium-magnesium carbonate) effervesces slowly in cold acid unless the mineral is very finely divided. If dolomite is suspected, place the sample in a container and warm it for 15 minutes after covering it with the acid solution.

Soluble Salts (saline, gypsiferous)

White incrustations that do not effervesce can be separated and checked for water solubility and taste (in uncontaminated soils). Chlorides, nitrates, and sulfates of sodium and potassium are very water soluble. **Chloride salts** can be tested by shaking a sample of soil in distilled water, placing about 10 mL of the supernatant solution (after the solids have settled) in a test tube, and adding a few drops of 5 percent sodium nitrate solution. Chlorides are indicated by the formation of a thick, milky precipitate of silver chloride. The formation of a heavy white precipitate after adding a 5 percent barium chloride solution to a separate supernatant sample indicates **sulfate** ions. SCS (1971, Section 19.6) provides further details for these tests.

Crystals of gypsum, which may occur as a white incrustation in voids, are rhombic plates, laths, or sometimes fibers. Gypsum crystals can be scratched with the fingernail, do not effervesce in acid, and are very slowly soluble in water.

Gypsum Acetone Test. Place 1 part soil and 10 parts water (by weight) in a small bottle. Seal the bottle and shake by hand at 15-minute intervals. Filter the extract through filter paper. Mix a 50-50 solution of the filtrate and acetone. The formation of a milky precipitate indicates the presence of gypsum. SCS (1971, Section 19.6) describes a somewhat more complex semi-quantitative test for gypsum using acetone which can be used in the field.

Iron Oxides (ferruginous)

Goethite and hematite commonly occur as segregated bodies in soils. Hematite is red; solid bodies, such as nodules or sheets may be dark brown or almost black but have a red streak if rubbed on a rough porcelain surface or a tough paper. Goethite bodies commonly are red but may be yellow or brown and are generally softer than hematite bodies. In an ignition test (see Section 3.1.3), hematite will show little color change; the duller colors of goethite will brighten when it changes to hematite. If gray, blue, or green materials turn red when ignited, ferrous iron is present. Table 3-11 provides additional information on color and occurrence of common iron oxide minerals in soil.

Reduced Iron

The presence of ferrous iron (Fe II), an indication that reducing conditions are present in the soil at least some of the time, can be tested using a neutral solution of $\alpha \alpha'$-dipyridyl dye (CAS No. 366-18-7) dissolved in 1 N ammonium acetate (Childs, 1981). The colorless dye solution should be applied to a freshly broken surface of a field-wet soil sample. The appearance of a red color on the dyed surface indicates presence of reduced iron. A negative response should not be considered definitive because the level of free iron may be below the sensitivity limit of the test or the soil may be an oxidized phase at the time of testing. The dye can result in false positive readings as a result of reduction of Fe(III) to Fe(II) in intense sunlight which then reacts with the dye.

Table 3-11 Color and Occurrence of Iron Oxide Minerals

Mineral	Color	Occurrence
Geothite	Yellowish-brown (7.5YR - 2.5Y)	Present in almost all soils
Hematite	Red (10R - 5YR)	Present mainly in well-drained soils
Maghemite*	Reddish brown (2.5 YR)	Well-drained soils of tropical areas; areas where soil has been burned (firepits, stumps)
Lepidocrocite	Orange (5YR-7.5YR, high chroma)	Noncalcareous soils with aquic moisture regimes
Ferrihydrite	Dark reddish-brown (5YR-7.5YR)	Concretions, placic horizons in Spodosols, brown scum in drainage ditches
Magnetite*	Black	A primary mineral inherited from igneous rock occurring as sand and silt-sized particles.

* Magnetic; other iron oxide minerals in table are not magnetic.

The dye itself is also sensitive to light and should also be kept out of sunlight, and solutions should be replaced periodically. Additional information about the dye, and safe handling procedures can be obtained from Sigma-Aldrich Corporation, 1001 West Saint Paul Ave., Milwaukee, WI (800/231-8327).

Manganese Oxide (manganiferous)

Black and very dark brown concretions and coatings on cleavage planes are likely to be the manganese oxide pyrolusite or a closely related mineral. It has a dark brown streak and is very soft, producing the streak even on paper. A procedure similar to the carbonate test using a dilute (5 percent) solution of hydrogen peroxide instead of HCl will result in the rapid evolution of small bubbles with a usually rapid consumption of the hydrogen peroxide if manganese oxides are present. See Section 3.3.1 for further guidance in differentiating possible reactions with organic matter.

SCS (1971, Section I7.2) discusses in more detail the identification of minerals and mineral groups in the field.

3.3.9 Fertility Potential

Soil fertility is the ability or status of a soil to supply water and nutrients necessary for plant growth. Inherent physical characteristics, such as soil structure and available water capacity, provide the basic elements of fertility potential, and are not easily modified. The nutrient status of a soil, on the other hand, can be improved by fertilization.

Nutrient status is evaluated by analyzing samples in the laboratory for nutrients essential for plant growth such as nitrogen, potassium, and phosphorus. Soil reaction (pH) is an important chemical parameter, because it strongly influences the availability of nutrients.

The key physical parameters affecting fertility potential are aeration and water availability, because plant roots require both air and a ready supply of moisture for optimum growth. These factors can be evaluated by observing soil properties such as soil texture (Section 3.1.2), depth to water table (Section 3.2.2), available water capacity (Section 3.2.3), and soil drainage class (Section 3.2.4). Deep, well-drained soils with a high available water capacity have the greatest fertility potential, provided nutrient status is favorable. Shallow soils with low available water capacity have low fertility potential even if nutrient status is favorable.

EPA's ESES defines the following soil fertility potential or status classes:

High: Nutrients necessary for plant growth readily supplied.

Moderate: Nutrients necessary for plant growth in moderate supply.

Low: Nutrients necessary for plant growth in low supply.

3.3.10 Soil Microbiota

Microorganisms in soil and ground water are now recognized as being of major importance in affecting the transformation and fate of many organic contaminants and heavy metals. The study of soil microbiota from soil samples requires carefully controlled laboratory conditions. The major concern in the field is that procedures used to collect soil samples do not allow contamination by microorganisms from other sources. Soil samples must be collected using sterilized tools and placed in sterilized containers. Samples of near-surface soils for microbiological study should be taken from a soil pit. Scoops or trowels used to collect samples from each soil horizon should go through usual decontamination procedures. As a final step, they should be heated with a blow torch and cooled by being stuck into the horizon to be sampled before soil material is dug out and placed into containers.

Samples for microbiological study taken from greater depths where oxygen is low require special aseptic handling procedures to prevent oxygen from harming anaerobic microorganisms. Leach et al. (1988) describe in detail procedures for collecting such samples, which involve preparing nitrogen-filled sampling containers in the laboratory, and using a field sampling glove box that has been purged with nitrogen gas to reduce the oxygen level below detectable limits.

EPA's ESES defines the following abundance classes for soil microorganisms (as number per gram of soil): **Abundant** ($> 1,000,000$), **common** ($10,000-1,000,000$); **few** ($100-10,000$), **none** (< 100).

3.4 Soil Contaminants

Visual identification of zones of soil contamination is sometimes possible when the contaminants are in an immiscible liquid phase or a solid phase. Iridescence of an oily phase in water may indicate contamination from synthetic organics. Solid phases should be described by color and consistency (granular, tarry, etc.). Possible contamination by heavy metals from artificial pigments is evident from bright colors. Toxic effects of contaminants on surface vegetation also may be evident. In any event, the presence of contaminants must be confirmed by laboratory analysis of samples.

Organic vapor detectors are relatively simple instruments that are used to identify the presence of volatile organics. The way in which readings are taken (distance from sample being checked, location in borehole, length of reading) should be recorded along with the readings themselves. Procedures should be performed consistently, and any departures from usual procedures noted.

Increasingly sophisticated field equipment in mobile laboratories is being used for onsite contaminant analysis. Use of such equipment is not covered in this guide.

CHAPTER 4
SOIL SAMPLING AND QUALITY ASSURANCE

Sample design, sample location, equipment and sampling methods, and quality assurance/quality control (QA/QC) procedures should all be determined before sampling begins and be recorded in the Soil Sampling and Quality Assurance Plans for the site. These documents will outline the specific procedures that will be required during field sampling. Carter (1993) provides information on collection and preparation of soil samples. Mason (1992) and Barth et al. (1989) provide detailed guidance on statistical aspects of sampling design and quality assurance. ASTM D4700-91 provides general guidance on selection of soil sampling equipment in the vadose zone, and ASTM C998-83 addresses surface sampling for radionuclides (Appendix E). ASTM D4220-89 and D5079-90 are useful references for methods for preserving and transporting soil and rock core samples. Carter (1993)

This chapter has two purposes: (1) provide information that may be useful if unforeseen conditions require modification of the procedures specified in the sampling plan, and (2) provide forms that may be useful in carrying out QA/QC procedures.

4.1 Changes in Soil Sampling Procedures

Soil description, use of field analytical equipment, and soil sampling should be conducted in a uniform and consistent manner, following procedures specified in the Soil Sampling and Quality Assurance Plans. If unforeseen conditions arise at the site during the field investigation and sampling that prevent carrying out the specified procedures, it may be necessary to develop alternative approaches at the site. Any such changes must be documented and approved (see Section 4.2).

One situation requiring departure from specified procedures is when soil conditions are unfavorable for the equipment being used to collect samples. If this situation occurs, the following tables may help identify alternative sampling tools:

> **Table 4-1** summarizes information on applications and limitations for use of 20 types of hand-operated soil sampling devices.
>
> **Table 4-2** summarizes information on applications and limitations for use of power-driven barrel and rotating core samplers (which obtain disturbed core samples).
>
> **Table 4-3** summarizes information on applications and limitations for use of 13 types of power-driven undisturbed core samplers.

Appendix B provides the names and addresses of two dozen manufacturers and distributors of manually operated and power-driven soil sampling equipment.

If the use of different sampling equipment requires altering the standard sampling protocol, the new protocol should be clearly specified. Appendices A.2 to A.4 provide some general soil handling and sampling protocols that can be used for guidance if revised protocols must be developed at the site. ASTM practices and methods for sampling using augers (D1452-80), thin-wall tubes (D1587-83), split barrel samplers (D1586-84), and ring-lined barrels (D3550-84) are useful references to have on hand in there has to be a change in sampling equipment type.

4.2 Quality Assurance/Quality Control

Any change in standard procedures for field collection of soil samples must be justified, described, and approved by the appropriate project personnel. Form 4-1 (Sample Alteration) can be used for this purpose. Multiple copies of this form should be available for use.

Soil sampling personnel should be aware that their work may be subject to a field audit to ensure that soil sampling and other QA/QC procedures are being followed. Form 4-2 contains a checklist of major items that should be covered in a field audit. Field personnel would do well to periodically review this checklist as a reminder of areas in which they would be held accountable in the event of a field audit.

The results of a field audit, or review of analytical results, may identify problems areas requiring corrective action. If the need for corrective action is identified during a field audit, it should be implemented immediately. Resampling may be required if analytical results fall outside the acceptable limits specified in the Quality Assurance Plan. Form 4-3 (Soil Sample Corrective Action Form) can be used to identify problem areas and specify measures required to correct the problems. When resampling is required, this form should be taken into the field and the specified procedures carefully followed.

Table 4-1. Summary of Hand-Held Soil Sampling Devices

Sampling Device	Applications	Limitations
Spoons and Scoops	Surface soil samples or the sides of pits or trenches	Limited to relatively shallow depths; disturbed samples
Shovels and Picks	A wide variety of soil conditions	Limited to relatively shallow depths
Augers*		
Screw Auger	Cohesive, soft, or hard soils or residue	Will not retain dry, loose, or granular, material
Standard Bucket Auger	General soil or residue	May not retain dry, loose, or granular material
Sand Bucket Auger	Bit designed to retain dry, loose, or granular material (silt, sand, and gravel)	Difficult to advance boring in cohesive soils
Mud Bucket Auger	Bit and bucket designed for wet silt and clay soil or residue	Will not retain dry, loose, or granular material
Dutch Auger	Designed specifically for wet, fibrous, or rooted soils (marshes)	
In-Situ Soil Recovery Auger	Collection of soil samples in reusable liners; closed top reduces contamination from caving sidewalls	Similar to standard bucket auger
Stony Soil Auger	Stony soils and asphalt	
Planer Auger	Clean out and flatten the bottom of predrilled holes	
Post-Hole/Iwan Auger	Cohesive, soft, or hard soils; readily available	Will not retain loose material

Table 4-1. (Continued)

Sampling Device	Applications	Limitations
Silage Auger	Silage pits and peat bogs	
Spiral Auger	Used to remove rock from auger holes so that borings can continues with other auger-type	
Split core auger	Auger with split core for easier recovery of sample; can be used with liner	
Tube Samplers**		
Soil Probe	Cohesive, soft soils or residue; representative samples in soft to medium cohesive soils and silts	Sampling depth generally limited to less than 1 meter
Thin-Walled Tubes	Cohesive, soft soils or residue; special tips for wet or dry soils available	Similar to Veihmeyer tube
Soil Recovery Probe	Similar to thin-wall tube; cores are collected in reusable liners, minimizing contact with the air	Similar to Veihmeyer tube
Veihmeyer Tube	Cohesive soils or residue to depth of 3 meters (maximum of 4.9 meters)	Difficult to drive into dense or hard material; will not retain dry, loose, or granular material; may be difficult to pull from ground
Geostick	Spot soil sampling and penetrometer tests	
Peat Sampler	Wet, fibrous, organic soils	

* Suitable for soils with limited coarse fragments; only the stony soil auger will work well in very gravelly soil.
** Not suitable for soils with coarse fragments.

Table 4-2. Summary of Major Types of Power-Driven Disturbed-Core Samplers

Tube Type	Applications	Limitations
Barrel Samplers		
Solid Barrel Sampler	Sand, silts, clays	Disturbed core, questionable recovery and quality below water table
Split Spoon Sampler	Disturbed samples from cohesive soils	Ineffective in cohesionless sands; not suitable for collection of samples for laboratory tests requiring undisturbed soil
Rotating Core		
Single Tube	Dense unconsolidated and consolidated formations	
Double-Tube	Friable, erodible, soluble or highly fractured formations	
CP Punch Core	Wireline system with various punch shoes; very effective in mixed formations where deep sampling is needed	

Source: Adapted from Rehm et al. (1985) and Aller et al. (1989).

Table 4-3. Summary of Major Types of Power-Driven Undisturbed-Core Samplers

Tube Type	Applications	Limitations
Thin-Wall Open Tube Samplers		
Shelby Tube	Undisturbed samples in cohesive soils, silt, and sand above water table	Ineffective in cohesionless sands or stony soil
Continuous Tube/Laskey Sampler	Same as Shelby tube, except longer barrel designed to operate inside the column of a hollow-stem auger	Same as Shelby tube; no blow counts taken
Thin-Wall Piston Samplers		
Internal Sleeve Piston Sampler	Collection of sample in heaving sands; used with hollow-stem auger with clamshell bit	Requires use of water or drilling mud for hydrostatic control; only one sample per borehole can be obtained
Wireline Piston Sampler	Undisturbed samples in cohesive soils and noncohesive sands; used with clam shell device on hollow-stem auger	In heaving sands only one sample per borehole can be collected because clamshell remains open after sampling
Fixed-Piston Sampler	Undisturbed samples in cohesive soils, silt, and sand above or below water table	Ineffective in cohesionless sands
Hydraulic Piston Sampler (Osterberg and others)	Similar to fixed-piston sampler	Not possible to limit the length of push or to determine amount of partial sampler penetration during push
Stationary Piston Sampler	Undisturbed samples in stiff, cohesive soils; representative samples in soft to medium cohesive soils, silts, and some sands	(Continued)

4-7

Table 4-3. (Continued)

Tube Type	Applications	Limitations
Gus Sampler	Similar to stationary piston sampler, except uses hydraulic action	
Free Piston Sampler	Similar to stationary piston sampler	Not suitable for cohesionless soils
Open Drive Sampler	Similar to stationary piston sampler	Not suitable for cohesionless soils

Specialized Thin-Wall (Section 2.4.5)

Pitcher Sampler	Undisturbed samples in hard, brittle, cohesive soils and cemented sands; representative samples in soft to medium cohesive soils, silts, and some sands; variable success with cohesionless soils	Frequently ineffective in cohesionless soils; require use of drilling fluid that may affect quality of sample
Denison Sampler	Undisturbed samples in stiff to hard cohesive soils, cemented sands, and soft rocks; variable success with cohesionless materials	Not suitable for undisturbed sampling of loose, cohesionless soils or soft cohesive soils; require use of drilling fluid that may affect quality of sample
Vicksburg Sampler	Similar to Shelby tube, but able to sample denser and coarser material	

Source: Adapted from Aller et al. (1989), Barrett et al. (1980), Rehm et al. (1985)

Form 4-1. Sample Alteration Form

Project Name and Number: _____

Material to Be Sampled: _____

Measurement Parameter: _____

Standard Procedure for Field Collection and Laboratory Analysis (cite references):

_____

_____

_____

Reason for Change in Field Procedure:

_____

_____

_____

_____

Variation for Field Procedure:

_____

_____

_____

_____

_____

Special Equipment, Materials, or Personnel Required:

_____

_____

_____

Initiator's Name: _____ Date: _____

Project Approval: _____ Date: _____

Laboratory Approval:_____ Date: _____

QA Officer/Reviewer: _____ Date: _____

Sample Control Center: _____ Date: _____

Form 4-2. Field Audit Checklist

Records to Inspect

_____ Chain-of-custody forms

_____ Analytical analysis request forms (if different from chain-of-custody forms)

_____ Sample tags

_____ Site description forms

_____ Log books

Sampling Procedures to Inspect

_____ Equipment

_____ Techniques

_____ Decontamination

_____ Collection of duplicate and field blank samples

_____ Security

_____ Sample storage and transportation

_____ Containers

_____ Contaminated waste storage and disposal

_____ Site description form entries

SOIL SAMPLING AND QUALITY ASSURANCE

Form 4-3. Sample Corrective Action Form

Project Name and Number: _____

Sample Data Involved: _____

Measurement Parameter(s): _____

Acceptable Data Range: _____

Problem Areas Requiring Corrective Action:

_____
_____
_____
_____
_____

Measures Required to Correct Problems:

_____
_____
_____
_____

Means of Detecting Problems and Verifying Correction:

_____
_____
_____
_____

Initiator's Name: _____ Date: _____

Project Approval: _____ Date: _____

Laboratory Approval:_____ Date: _____

QA Officer/Reviewer: _____ Date: _____

Sample Control Center: _____ Date: _____

4-11

REFERENCES

Aller, L., et al. 1989. Handbook of Suggested Practices for the Design and Installation of Ground-Water Monitoring Wells. EPA/600/4-89/034. [Published in 1989 by National Water Well Association, Dublin, OH in its NWWA/EPA series, 398 pp; published in March 1991 by U.S. EPA Center for Environmental Research Information, Cincinnati, OH, 221 pp]

Barrett, J. et al. 1980. Procedures Recommended for Overburden and Hydrologic Studies of Surface Mines. GTR-INT-71. U.S. Forest Service, Intermountain Forest and Experiment Station, Ogden, Ut, 106 pp.

Barth, D.S., B.J. Mason, T.H. Starks, and K.W. Brown. 1989. Soil Sampling Quality Assurance User's Guide, 2nd ed. EPA 600/8-89/046 (NTIS PB89-189864), 225+ pp.

Bingham, J.M. and E.J. Ciolkosz (eds.). 1993. Soil Color. Soil Science Society of America Special Publication No. 31, Madison, WI, 159 pp.

Blake, G.R. and K.H. Hartge. 1986. Bulk Density. In: Methods of Soil Analysis, Part 1, 2nd ed., A. Klute (ed.), Agronomy Monograph No. 9. American Society of Agronomy, Madison, WI, pp. 363-275.

Bradford, J.M. 1986. Penetrability. In: Methods of Soil Analysis, Part 1, 2nd ed., A. Klute (ed.), Agronomy Monograph No. 9. American Society of Agronomy, Madison, WI, pp. 463-478.

Breckenridge, R.P., J.R. Williams, and J.F. Keck. 1991. Characterizing Soils for Hazardous Waste Site Assessments. Ground-Water Issue Paper EPA/600/8-91/008. Available from CERI*.

Brown, R.H., A.A. Konoplyantsev, J. Ineson, and V.S. Kovalensky. 1983. Ground-Water Studies: An International Guide for Research and Practice. Studies and Reports in Hydrology No. 7. UNESCO, Paris. [Originally published in 1972, with supplements added in 1973, 1975, 1977, and 1983].

Brown, K.W., R.P. Breckinridge, and R.C. Rope. 1991. Soil Sampling Reference Field Methods. U.S. Fish and Wildlife Service Lands Contaminant Monitoring Operations Manual, Appendix J. Prepared by Center for Environmental Monitoring and Assessment, Idaho National Engineering Laboratory, Idaho Falls, ID, 83415. [Final publication pending revisions resulting from field testing of manual].

Cameron, R.E. 1991. Guide To Site and Soil Description for Hazardous Waste Sites, Vol. 1, Metals. EPA 600/4-91-029 (NTIS PB92-146158).

Cameron, R.E., G.B. Blank, and D.R. Gensel. 1966. Sampling and Handling of Desert Soils. NASA Technical Report No. 32-908. Jet Propulsion Laboratory, California Institute of Technology, 4800 Oak Grove Dr., Pasadena, CA, 91109.

Carter, M.R. (ed.). 1993. Soil Sampling and Methods of Analysis. Lewis Publishers, Boca Raton, FL, 528 pp.

Childs, C.W. 1981. Field Test for Ferrous Iron and Ferric-Organic Complexes (On Exchange Sites or in Water Soluble Forms) in Soils. Australian Journal of Soil Research 19:175-180.

Follmer, L.R., E.D. McKay, J.A. Lineback, and D.L. Gross. 1979. Wisconsinan, Sangamonian, and Illinoian Stratigraphy in Central Illinois. ISGS Guidebook 13, Appendix 3Illinois State Geological Survey, Champaign, IL.

Ford, P.J., P.J. Turina, and D.E. Seely. 1984. Characterization of Hazardous Waste Sites--A Methods Manual: Vol. II. Available Sampling Methods, 2nd ed. EPA 600/4-84-076 (NTIS PB85-521596).

Fritton, D.D. 1990. A Standard for Interpreting Soil Penetrometer Measurements. Soil Science 150(2):542-551.

Gee, G.W. amd J.W. Bauder. 1986. Particle-Size Analysis. In: Methods of Soil Analysis, Part 1, 2nd ed., A. Klute (ed.), Agronomy Monograph No. 9, American Society of Agronomy, Madison, WI, pp. 383-411.

Glocker, C.L. and L.A. Quandt (eds.). 1992. What's Up In Fragipans?, 2nd ed. Soil Survey Quality Assurance, National Soil Survey Center, Lincoln, NE, 61 pp.

Guthrie, R.L. and J.E. Witty. 1982. New Designations for Soil Horizons and Layers and the New Soil Survey Manual. Soil Sci. Soc. Am. J. 46:443-444.

Indiana Department of Environmental Management (IDEM). 1988. Requirements for Describing Unconsolidated Deposits (draft revised 11/18/88). IDEM, Indianapolis, IN.

Johnson, W.M., J.E. McClelland, S.B. McCaleb, R. Ulrich, W.G. Hoper, and T.B. Hutchings. 1960. Classification and Description of Soil Pores. Soil Science 89:319-321.

Langan, L.N. and D.A. Lammers. 1992. Definitive Criteria for Hydrologic Soil Groups. Soil Survey Horizons 32(3):69-77.

Leach, L.E., F.P. Beck, J.T. Wilson and D.H. Kampbell. 1988. Aseptic Subsurface Sampling Techniques for Hollow-Stem Auger Drilling. In: Proc. Third Nat. Outdoor Action Conf. on Aquifer Restoration, Ground Water Monitoring and Geophysical Methods. National Water Well Association, Dublin, OH, pp. 31-51.

Luxmoore, R.J. 1981. Micro-, Meso-, and Macroporosity of Soil. Soil Sci. Soc. Am. J. 45:671-672.

Lynn, W.C., W.E. McKinzie, and R.B. Grossman. 1974. Field Laboratory Tests for Characterization of Histosols. In: Histosols: Their Characterization, Use and Classification. Soils Science Society of America, Madison, WI, pp. 11-20.

Mason, B.J. 1992. Preparation of Soil Sampling Protocol: Techniques and Strategies. EPA/600/R-92/128 (NTIS PB92-220532). Environmental Monitoring Systems Laboratory, Las Vegas, NV, 89193-3478.

O'Neal, A.M. 1952. A Key for Evaluating Soil Permeability by Means of Certain Field Clues. Soil Sci. Soc. Am. Proc. 16:312-315.

Rehm, B.W., T.R. Stolzenburg, and D.G. Nichols. 1985. Field Measurement Methods for Hydrogeologic Investigations: A Critical Review of the Literature. EPRI EA-4301. Electric Power Research Institute, Palo Alto, CA.

Rhoades, J.D. and J.D. Oster. 1986. Solute Content. In: Methods of Soil Analysis, Part 1, 2nd ed., A. Klute (ed.), Agronomy Monograph No. 9. American Society of Agronomy, Madison, WI, pp. 985-1006.

Richards, L.A. (ed.). 1954. Diagnosis and Improvement of Saline and Alkali Soils. U.S. Department of Agriculture Handbook No. 60, 160 pp.

Richards, L.A., C.A. Bower, and M. Fireman. 1956. Tests for Salinity and Sodium Status of Soil and Irrigation Water. U.S. Department of Agriculture Circular 982.

Smith, G.D., F. Newhall, and L.H. Robinson. 1960. Soil-Temperature Regimes--Their Characterization and Predictability. SCS-TP-144. USDA Soil Conservation Service, 14 pp.

Soil Conservation Service (SCS). 1971. Handbook of Soil Survey Investigations Procedures. SCS, Washington, DC, 98 pp.

Soil Conservation Service (SCS). 1983. National Soils Handbook. [Available for inspection at SCS State, Area, and possibly County Offices. A review draft was released in September 1992]

Soil Conservation Service (SCS). 1984. Procedures for Collecting Soil Samples and Methods of Analysis for Soil Survey. Soil Survey Investigations Report No. 1. U.S. Government Printing Office.

Soil Conservation Service (SCS). 1990. Elementary Soil Engineering. In: Engineering Field Manual for Conservation Practices. SCS, Washington, DC, Chapter 4.

Soil Conservation Service (SCS). 1992. Draft National Soil Survey Interpretations Handbook. [Available for inspection at SCS State, Area, and possibly County Offices]

Soil Science Society of America. 1987. Glossary of Soil Science Terms. SSSA, Madison, Wisconsin.

Soil Survey Staff. 1975. Soil Taxonomy: A Basic System of Soil Classification for Making and Interpreting Soil Surveys. U.S. Department of Agriculture Agricultural Handbook No. 436.

Soil Survey Staff. 1992. Keys to Soil Taxonomy, 5th ed. SMSS Technical Monograph No. 19. Pocahontas Press, P.O. Drawer F, Blacksburg, VA, 24063-1020, 541 pp. ($20.00 plus $2.50 postage and handling; 5% discount for prepaid orders).

Soil Survey Staff. 1993. Examination and Description of Soils. In: Soil Survey Manual (new edition). Agricultural Handbook No. 18. Soil Conservation Service, Washington, DC, Chapter 3. [Note that this supersedes the 1951 Handbook by the same title, and 1962 supplement. Page proofs were being corrected at the time revisions to this guide were completed]

Taylor, S.A. and R.D. Jackson. 1986. Temperature. In: Methods of Soil Analysis, Part 1, 2nd ed., A. Klute (ed.), Agronomy Monograph No. 9. American Society of Agronomy, Madison, WI, pp. 927-940.

van Ee, J.J., L.J. Blume, and T.H. Starks. 1990. A Rationale for the Assessment of Errors in the Sampling of Soils. EPA/600/4-90/013. Environmental Monitoring Systems Laboratory, Las Vegas, NV, 89193-3478, 57 pp.

Vogel, W.G. 1987. A Manual for Training Reclamation Inspectors in the Fundamentals of Soils and Revegetation. Soil and Water Conservation Society, Ankeny, IA, 178 pp.

Witty, J.E. and E.G. Knox. 1989. Identification, Role in Soil Taxonomy, and Worldwide Distribution of Fragipans. In: Fragipans: Their Occurrence, Classification, and Genesis, N.E. Smeck and E.J. Ciolkosz (eds.), SSSA Sp. Pub. No. 24. Soil Science Society of America, Madison, WI, pp. 1-9.

APPENDIX A.1

GENERAL PROTOCOL FOR DESCRIPTION OF
SOIL CORES

Careful description of soil conditions at sampling locations can provide valuable information for interpreting soil analyses. Soil cores, which provide relatively undisturbed cross sections of the soil, are best for soil description. A few major features like texture, color (but not accurate description of mottling or variations in color), and potential zones of contamination, can be described from auger samples, but not much more.

At the outset, it should be decided whether soil descriptions will be made from the actual samples to be analyzed or from separate cores taken at the site. Describing actual samples has the advantage of allowing direct correlation of analyses with observed features, but will result in longer exposure of the sample to the air before it is placed in the sample container. This may not be desirable, even for samples taken for analysis of semivolatiles and metals. Table 1-1 contains an abbreviated list of suggested features to describe from soil samples.

Taking separate cores allows more leisurely and detailed observation of soil features. The soil can also be handled as necessary without concern about affecting its integrity for analyses. When equipment is being used just to describe soil features, decontamination procedures between locations also may be less rigorous, provided that there is no danger than contaminants with very low detection limits could be spread to uncontaminated areas.

The upper 1.5 to 2 meters, which have been affected by soil weathering processes, should receive the most careful attention for description because this is where the most complex features are likely to be encountered. Weathering may extend below 2 m in older landscapes formed in temperate to humid climates. Below the zone of weathering, the more simplified descriptions typical of geologic drill logs are appropriate. General procedures for describing both types of cores are outlined below.

Soil Cores (weathered zone, usually 1.5 to 2 m)

1. Near the location where the core is to be taken, spread a plastic sheet about 30 cm wide and 2 meters long on the ground, and place on it a fully extended carpenters rule or range pole marked with gradations that match the depth increments on the tube sampler to be used (i.e., in./ft or cm/meters).

2. Clear any litter away from the ground surface and take the first core. An open tube sampler that exposes most of the core when it is pulled out is easiest to use for this purpose. Remove the core and place it on the sheet at the zero end of the rule. If the presence of volatile contaminants is suspected, take the reading of the core as soon as it is brought to the surface with field instrumentation (photoionization or flame ionization detector). Also take a reading near the top of the core hole and record the measurements.

3. Repeat coring process trying to take equal increments (12 in. increments are usually possible in the upper 2 or 3 ft; 6 in. increments when the soil is very dry and in deeper, denser horizons), placing each core on the sheet at the appropriate depth interval until the desired maximum depth has been reached. If the presence of volatile contaminants is suspected, readings with field instruments should be made of each core as soon as it is brought to the surface, and near the top of the hole before taking the next increment.

4. If penetration is difficult in very dense or very dry soils, a weighted plastic mallet can be used to drive a sampler with a T-handle into the ground. If such conditions are typical, consider using samplers with specially designed weighted drivers.

5. If rock fragments prevent further penetration of the tube sampler before the desired depth has been obtained, an auger (screw or bucket) can be used. If the diameter of the auger is larger than that of the tube sampler (usually the case), discard

the soil material brought up by the auger for the depth increments already sampled.

6.   When the depth of interest is reached, pull up the auger at regular intervals and place soil by the rule on the sheet at the appropriate depth location. To prevent mixing of loose soil material from different depth increments, place material at chosen increments (e.g., 12 in.) on opposite sides of the rule.

7.   Once the complete soil is laid out on the sheet, visually examine the cores and place nails or some other kind of marker at the places where color changes indicate transitions between horizons.

8.   Carefully split the cores. A knife may be used to create a shallow groove, but the core should not be sliced all the way through because this will disturb structural features. Place one-half with the interior side facing up, and the other half with the interior side facing down (exposing the surface that was used for the initial visual inspection).

9.   Visually examine the interior face of the cores for transitions in structural units, texture, or other features. Adjust the locations of initially placed horizon markers on the sheet, if appropriate, and add additional markers for subhorizons, if required.

10.   Describe each soil horizon using Form 3-1. Refer to the appropriate sections of Chapter 3 for procedures and abbreviated codes for descriptors of soil parameters.

Cores Below Weathered Zone (Usually > 1.5 to 2 m)

Cores from greater depths can be described using essentially the same procedures as described above, except that the greater length requires placement of cores in holders where depth increments are side-by-side rather than end-on-end, and that generally fewer features are described. Most drillers and consultants have their own drill log forms. At a minimum, the following information should be recorded when describing cores below the weathering horizon:

1. Type of sample (split spoon, shelby tube, etc.)
2. Thickness driven/thickness recovered
3. Blow count (per 6 in.), if driven
4. Depth interval

Descriptions of depth intervals tend to be more abbreviated than near-surface soil profile descriptions and apply to regular depth intervals rather than transitions between horizons (although such transitions should be noted and described). Features should be described in a consistent sequence. The following features, when present should be described:

1. Texture (USDA and Unified estimated textures, coarse fragments)
2. Sorting and roundness
3. Moisture condition (moist, wet, dry, presence of water table)
4. Color and mottling
5. Consistency (rupture resistance, cementation)
6. Secondary porosity features
7. Sedimentary structures
8. Presence of organic matter
9. Effervescence in dilute, 10 percent cold HCl (calcareous parent material)
10. Visible presence of synthetic chemicals (oil, gasoline, solvents)
11. Reading from field instrumentation (photoionization or flame ionization detector)

APPENDIX A.2

GENERAL PROTOCOL FOR SOIL SAMPLE HANDLING AND PREPARATION

If questions arise in the field concerning sample handling and preparation procedures as specified in the Soil Sampling Plan for the site, this general protocol can be used. Any departures from procedures contained in the site Soil Sampling Plan should be documented and justified (see Form 4-1). The procedures described here generally apply to any type of soil sampling. They have been compiled primarily from procedures described in Brown et al. (1991). Specific procedures for different types of sampling tools are described in Appendices A.3 and A.4.

A.2.1 Soil Sample Collection Procedures for Volatiles (See, also, ASTM D4547-91, Appendix E)

1. Tube samplers are preferred when collecting for volatiles. Augers should be used only if soil conditions make collection of undisturbed cores impossible. Soil recovery probes and augers, with dedicated or reusable liners (see Table 4-1), will minimize contact of the sample with the atmosphere.

2. Place the first adequate grab sample, maintaining and handling the sample in as undisturbed a state as possible, in 40-mL septum vials or in a 1-L glass wide mouth bottle with a Teflon®-lined cap. *Do not mix or sieve soil samples.*

3. Ensure the 40-mL containers are filled to the top to minimize volatile loss. Secure the cap tightly.

4. Examine the hole from which the sample was taken with an organic vapor instrument after each sample increment. Record any instrument readings.

5. Label and tag sample containers, and record appropriate data on soil sample data sheets (depth, location, etc.).

6.   Place glass sample containers in sealable plastic bags, if required, and place containers in iced shipping container. Samples should be cooled to 4°C as soon as possible.

7.   Complete chain-of-custody forms and ship as soon as possible to minimize sample holding time (see Table A-1 for maximum holding times for various constituents.

8.   Follow required decontamination and disposal procedures (see A.2.3)

A.2.2   Soil Sample Collection and Mixing Procedures for Semivolatiles and Metals

1.   Collect samples.

2.   If required, composite the grab samples, or use discrete grab samples.

3.   If possible, screen the soils in the field through a precleaned O-mesh (No. 10, 2 mm) stainless steel screen for semivolatiles, or Teflon®-lined screen for metals (some metals in stainless steel could contaminate the sample).

4.   Mix the sample in a stainless steel, aluminum (not suitable when testing for Al), or glass mixing container using the appropriate tool (stainless steel spoon, trowel, or pestle).

5.   After thorough mixing, place the sample in the middle of a relatively inexpensive 1-m square piece of suitable plastic, canvas, or rubber sheeting.

6.   Roll the sample backward and forward on the sheet while alternately lifting and releasing opposite sides or corners of the sheet.

7.   After thorough mixing, spread the soil out evenly on the sheet with a stainless steel spoon, trowel, spatula, or large knife.

8.   Take sample container and check that a Teflon® liner is present in the cap, if required (see Table A-1 for recommended sample containers for different contaminants).

A-6

9.	Divide the sample into quarters, and take samples from each quarter in a consecutive manner until appropriate sampling volume is collected for each required container. Separate sample containers would be required for semivolatiles, metals, duplicate samples, triplicate samples (split), and spiked samples.

10.	Secure the cap tightly. The chemical preservation of solids is generally not recommended.

11.	Label and tag sample containers, and record appropriate data on soil sample data sheets (depth, location, other observations).

12.	Place glass sample containers in sealable plastic bags, if required, and place containers in iced shipping container. Samples should be cooled to 4°C as soon as possible.

13.	Complete chain-of-custody forms and ship as soon as possible to minimize sample holding time (see Table A-1 for maximum holding times for various constituents). Scheduled arrival time at the analytical laboratory should give as much holding time as possible for scheduling of sample analyses.

14.	Follow required decontamination and disposal procedures (see A.2.3)

A.2.3 Equipment Decontamination/Disposal

Decontamination procedures may vary from state to state and site to site. Detailed procedures should be specified in the Soil Sampling Plan, such as those contained in ASTM D5088-90 (Appendix E). A very general procedure is outlined here:

1.	Any disposable solid contaminated equipment (plastic sheets, screens, etc.) should be placed in plastic bags for temporary storage and sealed in metal barrels for final transport/disposal.

2.	Reusable equipment should be washed and rinsed using decontamination procedures specified in the Soil Sampling Plan.

3.   Collect swipes and decontamination blanks, if required, to evaluate the possibility of cross-contamination.

A.2.4 Air Drying

Samples collected for chemical analysis in the laboratory do not normally need to be air dried. However, air drying may be desired in the field for evaluation of soil physical and hydrologic properties. In some instances, air drying of contaminated samples involving semivolatiles and metals may be desired before sending samples to the analytical laboratory.

1.   Weigh sample and record weight if percent moisture is required.

2.   Spread out the soil sample on a stainless steel sheet and allow to air dry. This may take 3 days or more. If samples are to be analyzed for possible contaminants, samples should be placed so as to prevent possible cross-contamination. If they are to be analyzed for microbiological activity, samples should be placed in containers through which filtered air can be passed.

3.   When dry, weigh and record weight if percent moisture is required.

4.   Break up soil aggregates and pull apart vegetation and root mat, if present. Weigh nonsoil vegetation fraction, and archive or discard, as required.

5.   Remove large rocks and weigh. Archive for possible analysis.

6.   Crush the entire soil sample with a rolling pin, stainless steel spoon, or some similar tool. Blend with stainless steel spoon for 30 minutes.

7. Sieve through an O-mesh (No. 10, 2 mm) screen. Any type of screen is acceptable, if soils are not contaminated. Use disposable stainless steel (semivolatile contamination) or Teflon® (metal contamination) if soil samples are contaminated and the chemical integrity of the sample must be maintained.

8. Spread out the sample, mark off quarters, and take sample from each quarter in a consecutive manner until appropriate sample volume is collected. Archive remaining sample for future analysis, if needed.

9. When ready for shipment to the analytical laboratory, shake the sample to mix thoroughly.

10. Follow required decontamination and disposal procedures (see A.2.3).

Table A-1. EPA Recommended Sampling Containers, Preservation Requirements, and Holding Times for Soil Samples

Contaminant	Container* Preservation**	Holding Time***
Acidity	P,G	14 days
Alkalinity	P,G	14 days
Ammonia	P,G	28 days
Sulfate	P,G	28 days
Sulfide	P,G	28 days
Sulfite	P,G	48 hours
Nitrate	P,G	48 hours
Nitrate-Nitrite	P,G	28 days
Nitrite	P,G	48 hours
Oil and Grease	G	28 days
Organic Carbon	P,G	28 days
Metals		
Chromium VI	P,G	48 hours
Mercury	P,G	28 days
Other Metals	P,G	6 months
Cyanide	P,G	28 days
Organic Compounds		

Extractables
Including Phthalates,
  Nitrosamines, Organo-
chlorine Pesticides,
  PCBs, Nitroaromatics
Isophorone, Polynuclear
Aromatic Hydrocarbons,
Haloethers, Chlorinated
Hydrocarbons, and TCDD
	G, Teflon®-
	lined cap
		7 days
		until
		extraction
		30 days
		after
		extraction

Extractables (Phenols)	G, Teflon®- lined cap	7 days until extraction 30 days after

Contaminant	Container* Preservation**	Holding Time***
		extraction
Purgeables		
Halocarbons and Aromatics	G, Teflon®-lined septum	14 days
Acrolein and Acrylonitrate	G, Teflon®-lined septum	3 days
Orthophosphate	P,G	48 hours
Pesticides	G, Teflon®-lined cap	7 days until extraction 30 days after extraction
Phenols	G	28 days
Phosphorus	G	48 hours
Phosphorus, Total	P,G	28 days
Chlorinated Organic Compounds	G, Teflon®-lined cap	7 days

*   P = polyethylene, G = glass.

**   All samples cooled to 4°C. Sample preservation should be performed immediately upon sample collection. For composite samples, each aliquot should be preserved at the time of collection. When impossible to preserve each aliquot, then samples may be preserved by maintaining at 4°C until compositing and sample splitting is completed.

*** Samples should be analyzed as soon as possible after collection. The times listed are the maximum times that samples may be held before analysis and still be considered valid. Samples may be held for longer periods only if the analytical laboratory has data on file to show that the specific types of samples under study are stable for the longer time.

Source: Barth et al. (1989).

APPENDIX A.3

GENERAL PROTOCOL FOR SOIL SAMPLING
WITH A SPADE AND SCOOP

The simplest and most direct method of collecting soil samples for subsequent analysis is with a spade and scoop. A normal lawn or garden spade can be used to remove the top cover of soil to the required depth, and then a smaller stainless steel scoop can be used to collect the sample.

This method can be used in most soil types, but is limited to sampling the near surface. Samples from depths greater than 50 cm become extremely labor intensive in most soil types. Very accurate, representative samples can be collected with this procedure depending on the care and precision demonstrated by the sampler. A flat, pointed mason trowel can be used to cut a block of soil when relatively undisturbed samples are desired. A stainless steel scoop or lab spoon will suffice in most other applications. Chrome-plated digging instruments, common with garden implements such as potting trowels, should be avoided.

Procedure   (drawn from Ford et al., 1984)

1.   Clear the area to be sampled of any surface debris (twigs, rocks, litter). It may be advisable to remove the first 8 to 15 cm of surface soil for an area approximately 15 cm in radius around the drilling location to prevent near-surface soil particles from falling down the hole.

2.   Carefully remove the top layer of soil to the desired sample depth with a precleaned spade.

3.   Using a precleaned stainless steel scoop or trowel, remove and discard a thin layer of soil from the area which came in contact with the shovel.

4.   Collect and handle sample using procedures described in A.2.1 (Soil Sample Collection Procedures for Volatiles) and A.2.2 (Soil Sample Collection and Mixing Procedures for Semivolatiles and Metals).

APPENDIX A.4

GENERAL PROTOCOL FOR SOIL SAMPLING WITH AUGERS AND THIN-WALL TUBE SAMPLERS

Hand-held augers and thin-wall tube samplers can be used separately or in combination. Where rocky soils do not limit the use of tube samplers, a combination of augers to remove soil material to the depth of interest and tube samplers for actual sample collection allows the most precise control of sample collection. Depths to 2 meters can be readily sampled and up to 6 meters where conditions are favorable. Tables 4-1 and 4-3 provide information on the advantages and disadvantages of different types of augers and tube samplers for sampling under different soil conditions.

The recently developed in situ soil recovery auger and probe allow collection of samples in dedicated or reusable liners that reduce cross-contamination of samples and minimize contact with the atmosphere (see Table 4-1, and Appendix B).

Specific sampling tools may require slightly different handling methods. For example, if sampling devices and drill rod extensions do not have quick connect fittings, crescent or pipe wrenches may be required to change equipment configuration. The procedure described below is for hand-held equipment. Procedures for power-driven augers or tube samplers are essentially the same (drawn from Ford et al., 1983, and Brown et al., 1991).

1. Attach the auger bit to a drill rod extension and further attach the "T" handle to the drill rod.

2. Clear the area to be sampled of any surface debris (twigs, rocks, litter). It may be advisable to remove the first 8 to 15 cm of surface soil for an area approximately 15 cm in radius around the drilling location to prevent near-surface soil particles from falling down the hole.

3.   Begin drilling, periodically removing accumulated soils. This prevents accidentally brushing loose material back down the borehole when removing the auger or adding drill rods.

4.   After reaching the desired depth, slowly and carefully remove auger from boring. When sampling directly from auger, collect sample after auger is removed from boring. Discard the upper portion of the sample, which may contain soil that has fallen to the bottom of the hole from the sidewalls. Proceed to sample handling and mixing procedures (see A.2.1 and A.2.2).

5.   If taking a core sample, remove auger tip from drill rods and replace with a precleaned thin-wall tube sampler. Install proper cutting tip. (An optional step is to first replace the auger tip with a planer auger to clean out and flatten the bottom of the hole before using the thin-wall tube sampler.)

6.   Carefully lower corer down borehole. Gradually force corer into soil. Care should be taken to avoid scraping the borehole sides. Hammering of the drill rods to facilitate coring should be avoided, as the vibrations may cause the bore walls to collapse.

7.   Remove corer and unscrew drill rods.

8.   Remove core from device (this may require removing cutting tip) and discard top of core (approximately 2.5 cm), to eliminate soil that may have fallen down from higher horizons.

9.   Handle sample using procedures described in A.2.1 and A.2.2.

APPENDIX B

Appendix B. Manufacturers and Distributors of Soil Sampling Equipment

Supplier	Types of Samplers
Acker Drill Company* P.O. Box 830 Scranton, PA 18501 800/752-2537	Manual samplers Power-driven samplers Continuous sampling systems
Art's Manufacturing and Supply (AMS) 105 Harrison American Falls, ID 83211 1/800/635-7330	Manual samplers** In-situ soil recovery auger and probe Planer auger Split core sampler
Ben Meadows Company, Inc. P.O. Box 80549 Atlanta (Chamblee), GA 30366 800/241-6401	Manual samplers** Clinometers Munsell Color Charts
Christensen Boyles Corporation Products Division 4446 West 1730 South P.O. Box 30777 Salt Lake City, UT 84130 800/453-8418	Manual samplers Geostick Power-driven samplers Laskey continuous sampler GUS piston sampler
C.F.E. Equipment 9 South Peru Street Plattsburgh, NY 12901 800/665-6794	Manual samplers** Stony soil auger
Clements Associates, Inc. RR 1 Box 186 Newton, IA 50208-9990 800/247-6630	Manual samplers**
Drillers Service, Inc. Environmental Products Division 1972 Highland Ave., NE P.O. Drawer 1407 Hickory, NC 28603 800/334-2308	Manual samplers** Power-driven samplers

Supplier	Types of Samplers
Forestry Suppliers P.O. Box 8397 Jackson, MS 39284-8397 800/647-5368	Manual samplers** Clinometer Munsell Color Charts
Geoprobe Systems 607 Barney St. Salina, KS 67401 713/825-1842	Power-driven samplers
Giddings Machine Company 401 Pine Street P.O. Box 2024 Fort Collins, CO 80522 303/482-5586	Power-driven samplers
Gilson Company, Inc. P.O. Box 677 Worthington, OH 43085-0677 800/444-1508	Manual samplers Pocket penetrometer Sieves
Hansen Machine Works 1628 North C Street Sacramento, CA 95814 916/443-7755	Veihmeyer probe
HAZCO Services, Inc. 2006 Springboro West Dayton, OH 45439 800/332-0435	Manual soil samplers** Waste samplers
Longyear Company 2340 West 1700 South Salt Lake City, UT 84104 801/972-6430	Power-driven samplers
Mobile Drilling Company 3807 Madison Ave. Indianapolis, IN 46227 800/766-3745	Power-driven samplers Continuous sampling systems

Supplier	Types of Samplers
Oakfield Apparatus Company P.O. Box 65 Oakfield, WI 53065 414/583-4114	Manual samplers
Penndrill Manufacturing Div. Pennsylvania Drilling Co. 500 Thompsen Ave. McKees Rocks, PA 15136 800/245-4420; 412/771-2110	Power-driven samplers
Solinst 515 Main St. Glen Williams Ontario L7G 3S9 416/873-2255	Power-driven samplers Thin-wall piston sampler
Soilmoisture Equipment Corp. P.O. Box 30025 Santa Barbara, CA 93105 805/964-3525	Manual samplers
Soiltest Products Division ELE International, Inc. P.O. Box 8004 Lake Bluff, IL 60044 800/323-1242	Manual samplers Power-driven samplers Penetrometers
Southern Iowa Manufacturing Drilling Products Division P.O. Box 448 Osceola, IA 50213 800/338-9925	Power-driven samplers
Wheaton 1301 North 10th Street Millville, NY 08332-9854 800/225-1437	Manual soil samplers Waste samplers

* Acker Drill Company was purchased by Christensen Boyles Corporation in 1992.
** Liners available for collection of samples with auger or probes that minimize cross contamination and contact of sample with air.

APPENDIX C

GUIDE FOR SELECTION OF PARAMETERS AND METHODS FOR CHARACTERIZATION OF CONTAMINATED SOILS

This appendix includes five checklists containing more than 100 soil and contaminant physical, chemical, and biological properties that may be of interest at a contaminated site:

1. Site characterization parameters (Checklist C-1).

2. Soil physical parameters (Checklist C-2).

3. Soil hydrologic parameters (Checklist C-3).

4. Soil chemical and biological parameters (Checklist C-4).

5. Contaminant chemical and biological parameters (Checklist C-5).

A set of five tables that parallel the checklists provides summary information on: (1) each parameters significance in relation to characterization and remediation of contaminated sites, (2) identification of field, laboratory, and "lookup" methods for measuring or characterizing the parameter, and references where more detailed information can be found about specific methods.

The term "lookup" method is used broadly here to include ways of determining or estimating a parameter value without any actual field or laboratory measurements, or from other measured soil properties. Lookup methods are the primary source of information for many basic physical and chemical properties of contaminants (Checklist C-5). Soil physical, hydrologic and chemical properties vary too much spatially to allow accurate estimation, but lookup methods may be useful during preliminary site characterization and modeling for identifying parameters of special significance in transport and fate of contaminants. This, in turn can guide selection of field and laboratory methods for more accurate, site-specific measurement of critical parameters. For example, field and laboratory measurements of soil

C-1

hydrologic properties are relatively complicated and time-consuming, but can be estimated relatively quickly based on other observed properties (see Table 3-9 for estimation of hydraulic conductivity from soil texture and structure).

Use of SCS Soil Data

A great many soil physical, hydrologic, and chemical properties can be inferred from a soil survey conducted in accordance with methods and procedures of SCS's National Cooperative Soil Survey Program. The basic taxonomic unit in SCS's soil taxonomy is the **soil series**. There are around 20,000 recognized soil series in the United States. Official soil series descriptions are maintained on a two-page form that contains the following information:

> Taxonomic Class
> Typical soil profile description
> Range of characteristics
> Competing series
> Geographic setting
> Geographically associated soils
> Drainage and permeability
> Use and vegetation
> Distribution and extent
> Location and year series was established
> Remarks
> Availability of additional data

SCS also maintains a computerized database of **soil interpretation records** (Form SCS-SOI-5) for each soil series that include the following estimated soil properties for each major horizon:

> Texture class (USDA, Unified, and AASHTO)
> Particle size distribution
> Liquid limit
> Plasticity index
> Moist bulk density
> Permeability (in/hr)
> Available water capacity (in/in)
> Soil reaction (pH)
> Salinity (mmhos/cm)
> Sodium adsorption ratio (SAR)

Cation exchange capacity (CEC)
Calcium carbonate (%)
Gypsum (%)
Organic matter (%)
Shrink-swell potential
Erosion factors (K,T)
Wind erodability group/index
Corrosivity (steel and concrete)
Flooding (frequency, duration, months)
High water table (depth, kind, months)
Cemented pan (depth, hardness)
Bedrock (depth, hardness)
Subsidence (initial, total)
Hydrologic group
Potential frost action

If a published SCS soil survey is available for a site of interest, information in the forms mentioned above will be contained in the report, but scattered in different locations. It is probably useful to obtain the individual soil series descriptions and interpretation records (often available from the SCS State Office as a four-page handout) as a convenient consolidated reference for each soil series of interest. However, this information should be checked against data in the published soil survey, since the soil survey often will have additional data that is specific to the county in question.

If available, a published SCS county soil survey, typically ranging from 1:15,840 to 1:20,000 mapped on a airphoto base, should be the starting point for any site-soil investigation. A review of the soil series descriptions and related information will provide a good idea of the kinds of soil and hydrologic conditions that are likely to be encountered at the site. Since the scale of SCS soil surveys may result in areas of contrasting soils as large as 3 or 4 acres being included in a map unit, detailed site investigations will usually result in some modifications of soil boundaries.

Using the Checklists

If vadose zone or ground water computer modeling is planned for a site, the input parameters required for the model or models to be used provide a good starting point for filling in the checklists in this appendix. The number of parameters that are checked for a particular site will depend on the objectives of the investigation. For example, soil engineering properties are not especially important in an investigation to assess the extent, direction and rate of movement of contaminants, but are essential when designing waste disposal facilities or corrective action programs.

The type of contaminant will also be a factor in selecting parameters to be investigated. Contamination with volatile organic compounds requires some knowledge of soil volatilization parameters (Checklist C-3), whereas this information is not required when heavy metals are the only contaminants.

A suggested procedure for using the checklists in this appendix is a follows:

1. Obtain available SCS soil maps, soil series descriptions and soil interpretation records for the site and check the "lookup" column for each parameter covered by this material.

2. Prepare a list of input parameters required for any computer models that are planned, or being considered for use at the site. Locate the parameters from this list on the checklists and check **in pencil** all columns for which a method is indicated.

3. Go through the checklists a third time and check **in pencil** any other parameters that appear to be necessary for characterizing the site in a way that will accomplish the objectives of the investigation.

4. Review each parameter that has been checked. If both field and laboratory methods are available, refer to the appropriate parameter/method table to determine which method(s) to use. In general, field methods for measurement of soil hydrologic properties are preferable to laboratory methods because they include a larger volume of soil material. Laboratory methods will require collection of samples in the field. Laboratory measurement of most physical and hydrologic parameters requires collection of undisturbed soil cores (see Tables 4-1 and 4-3). This is not as critical for samples collected for chemical analysis. The penciled checks for methods that are **not** ultimately selected (including lookup methods for parameters that are not relevant) should be erased, and the remaining checks marked over with ink.

5. The completed checklists will identify which soil physical and chemical test procedures to mark in Form 1-1, which serves as a quick reference for locating field tests described in this guide. A list of additional equipment for more complex field tests (such as measurement of soil moisture, matric potential, and hydraulic conductivity) should be added to Form 1-2 (Soil Description/Sampling Equipment Checklist).

6. As field investigations proceed, the completed checklists from this appendix should be reviewed to determine if new information about the site requires collection of data on additional parameters.

Using the Parameter/Methods Tables

The parameters in Checklists C-1 through C-5 are listed in numerical sequence with letter designations of parameters that fall within a larger category. For example, the eleven soil engineering parameters are designated as numbers 13a through 13k. This numbering makes it easier to locate the parameters in the five parameter/method tables that follow the checklists. Descriptions of methods in these tables are necessarily brief, often limited to a listing of ways in which a parameter can be measured. ASTM standards applicable to the method are identified with ASTM's alphanumeric designation. Appendix E provides the full title and volume of the Annual Book of ASTM Standards in which a standard is located.

Other major references that provide guidance on specific methods may also be provided. All other references cited in these tables are contained at the end of this appendix.

The U.S. Environmental Protection Agency's two-volume guide **Subsurface Characterization and Monitoring Techniques** (Boulding 1993), available from EPA's Center for Environmental Research Information (ORD Publications, P.O. Box 19963, Cincinnati, OH 45219-0963; 513/569-7562) is recommended as a companion to the tables in this appendix. This guide contains one-to two-page summary sheets on more than 270 specific techniques for characterization and monitoring of soil, the vadose zone and ground water. Each summary provides information on uses at contaminated sites, a description of the method and advantages and disadvantages of the method.

Chemical Data Sources

Checklist C-5 contains 29 contaminant chemical and biological parameters that may be of importance in evaluating degree of hazard, fate and potential for transport at a site. Most contaminant hazardous properties and general chemical properties and volatilization characteristics are independent of site conditions and can be obtained from chemical reference sources. Contaminant sorption/retention and biodegradation parameters may be highly site specific, but data from other sites and laboratory tests may be useful for preliminary evaluation of transport potential. As many reference sources as possible should be consulted when collecting information on known or suspected contaminants at a site. Table C-6 identifies more than 50 major reference sources in three major categories: (1) general chemical references, (2) hazardous chemical references, and (3) references providing chemical fate data.

Checklist C-1. Checklist of Site Characterization Parameters for Field, Laboratory and Calculation/Look-Up Methods (see, also, Table C-1)

Parameter	Field	Lab	Lookup

1. Water Budget Parameters

a. Precipitation   _____   _____
b. Infiltration*   _____   _____
c. Evaporation   _____   _____
d. Evapotranspiration   _____   _____   _____
e. Surface Runoff*   _____   _____

2. Other Climate Parameters

a. Air temperature*   _____   _____
b. Wind speed/direction*   _____   _____
c. Humidity   _____   _____
d. Insolation   _____   _____

3. Geomorphology

a. Slope gradient/length*   _____   _____
b. Slope form/landscape position   _____   _____

4. Erodiblity

a. Water Erosion*   _____   _____
b. Wind Erosion   _____   _____

5. Surface Hydrology

a. Surface streams   _____   _____
b. Flood frequency/duration   _____   _____
c. Water Bodies _____   _____

6. Biota (see 21.a-d for soil microbiota)

a. Vegetation*   _____   _____   _____
b. Macrofauna and Mesofauna*   _____   _____

* Included in EPA/600/4-91/029 (Cameron, 1991)

Checklist C-2. **Checklist of Soil Physical Parameters for Field, Laboratory and Calculation/Look-Up Methods (see, also, Table C-2).**

Parameter	Field	Lab	Lookup
7. Horizons*	____	____	____
8. Texture*	____	____	____
9. Color*	____	____	____
10. Porosity*	____	____	
11. Zones of Increased Porosity/Permeability			
a. Soil Structure/Cracks*	____	____	____
b. Roots*	____	____	
c. Lateral features*	____	____	
d. Sedimentary features	____	____	
12. Zones of Reduced Porosity/Permeability			
a. Genetic horizons	____	____	____
b. Rupture resistance*	____	____	
c. Bulk density*	____	____	____
d. Restrictive layers	____		
13. Soil Engineering			
a. ASTM (Unified) texture	____	____	____
b. Atterberg limits	____	____	____
c. Shear strength	____	____	
d. Shrink-swell ____	____	____	
e. Corrosivity*	____	____	____
f. Compaction*	____	____	
g. Compressibility		____	____
h. Bearing capacity		____	
i. Erosion resistance	____	____	____
j. Clay dispersivity	____	____	
k. Frost heave	____		____

* Included in EPA/600/4-91/029 (Cameron, 1991)

Checklist C-3. Checklist of Soil Hydrologic and Air Parameters for Field, Laboratory and Calculation/Look-Up Methods (see, also, Table C-3).

Parameter	Field	Lab	Lookup
14. Soil Water State			
a. Moisture content*	____	____	____
b. Water potential	____	____	____
c. Specific retention	____	____	____
d. Available water capacity	____	____	____
e. Moisture regime*			____
15. Internal Free Water (Saturated Zone)			
Depth/Thickness/Duration	____		____
16. Permeability/Hydraulic Conductivity			
a. Saturated*	____	____	____
b. Unsaturated* ____	____	____	
17. Contaminant Transport in Water			
a. Velocity	____	____	____
b. Water/solute flux	____	____	____
c. Dispersivity	____	____	____
18. Volatilization (see, also 20a to 20e)			
a. Air permeability	____	____	____
b. Gas diffusivity	____	____	
c. Gas flux	____	____	
d. Air temperature (see 2.a)*	____		____
e. Wind speed (see 2.b)*	____		____
19. Soil Temperature			
a. Soil temperature*	____		____
b. Soil temperature regime*	____		____

* Included in EPA/600/4-91/029 (Cameron, 1991)

Checklist C-4. **Checklist of Soil Chemical and Biological Parameters for Field, Laboratory and Calculation/Look-Up Methods (see, also, Table C-4).**

Parameter	Field	Lab	Lookup
20. Soil Chemistry			
a. Organic carbon/matter*	___	___	___
b. Odor*	___		
c. Cation exchange capacity*	___	___	___
d. Soil pH*	___	___	___
e. Soil oxygen	___	___	___
f. Redox potential (Eh)*	___	___	___
g. Redox couple ratios (waste/soil system)	___	___	
h. Clay Mineralogy*	___	___	___
i. Other Mineralogy	___	___	___
j. Salinity (EC)*	___	___	___
k. Sodicity (SAR)	___	___	___
l. Major cations	___	___	
m. Major anions	___	___	
n. Fertility potential*	___	___	___
21. Soil Microbiota*			
a. Enumeration*	___	___	
b. Metabolism	___	___	
c. Nutrients	___	___	
d. Activity/kinetics	___	___	
22. Soil Pollution Situation*			
a. Type	___	___	___
b. Concentration	___	___	
c. Odor*	___		
d. Depth	___		
e. Volume	___		
f. Date(s) of contamination	___		___

* Included in EPA/600/4-91/029 (Cameron, 1991)

Checklist C-5.   Checklist of Contaminant Chemical and Biological
                 Parameters for Field, Laboratory and
                 Calculation/Look-Up Methods (see, also, Table C-5).

Parameter	Field	Lab	Lookup
23. Contaminant Hazardous Properties			
a. Toxicity	___		___
b. Reactivity	___		___
c. Corrosivity	___		___
d. Ignitibility	___		___
24. Contaminant General Chemical Properties			
a. Chemical class			___
b. Molecular weight/structure			___
c. Melting point/boiling point			___
d. Specific gravity/density		___	___
e. Water solubility/miscibility		___	___
f. Speciation		___	___
g. Dielectric constant			___
25. Contaminant Chemical Reactivity			
a. Acid/base		___	___
b. Oxidation/reduction		___	___
c. Complexation		___	___
d. Hydrolysis		___	___
e. Catalysis		___	___
f. Precipitation/dissolution		___	___
g. Polymerization		___	___
26. Contaminant Volatilization Parameters			
a. Henry's Constant (Kh)			___
b. Vapor pressure		___	___
c. Vaporization temp./solubility (see 24.c and e)			
d. Koc sorption (gaseous phase)		___	___
27. Soil Contaminant Sorption/Retention			
a. Koc	___	___	___
b. Linear Kd	___	___	___
c. Nonlinear sorption constants	___	___	
d. Octanol-water (Kow)		___	___
e. Residual saturation (Ko)	___	___	
28. Soil Contaminant Degradation			
a. Half-life/rate constant		___	___
b. Biodegradability	___	___	___
c. Degradation products	___	___	___

* Included in EPA/600/4-91/029 (Cameron, 1991)

Table C-1. Site Characterization Parameters of Interest and Applicable Field, Laboratory, and Calculation/Lookup Methods

Site Parameter/Significance	Methods/References
1. Water Budget. Since long term data are often needed for water budget analysis it is not always feasible to make measurements at contaminated sites.	
a. Precipitation. Input data to contaminant migration models may include monthly precipitation, storm event precipitation, periods between storms, and snowfall.	**Field:** Sacramento gage (accumulated precipitation, manual recording), weighing gage (continuous measurement, mechanical recording), or tipping-bucket gage (continuous measurement with electronic recording)—Boulding (1993, Chapter 8). **Laboratory:** — . **Lookup:** Precipitation data collected by National Weather Service or other source for area near site (Hatch, 1988). Interpolation using published maps of precipitation data.
b. Infiltration is the rate at which water enters the soil. Affects the amount of precipitation that enters the soil and how much runs off. Low infiltration required for caps.	**Field:** Cylinder infiltrometer (ASTM D3385-88/D5093-90; Bouwer, 1986), test basins (U.S. Army Corps of Engineers (1980), sprinkler infiltrometer (Peterson and Bubenzer (1986); **Laboratory:** —. **Lookup:** Methods available for estimating infiltration of small watersheds (Dunne and Leopold, 1978), large watersheds (Musgrave and Holtan (1964). Infiltration equations can be used with field-measured or literature estimates (Green and Ampt, 1911; Philip 1957).
c. Evaporation is the rate at which surface water returns to the atmosphere as water vapor. Required for modeling of sites where surface water bodies exist.	**Field:** Class-A Pan evaporation from surface of free liquid (NWS 1972; USGS, 1982). **Laboratory:** — . **Lookup:** Interpolation using maps showing average evaporation; empirical and physically-based evaporation equations (Boulding, 1993, Sections 8.4.1 and 8.4.2).

Table C-1. (Continued)

Site Parameter/Significance	Methods/References

1d. Evapotranspiration is the rate at which water evaporates from soil and plants transpire water. Essential for water budget modeling.

Field: Water balance methods include: lysimeters (nonweighable, weighing, hydraulic/floating, monolith/soil block), soil moisture monitoring, evaporimeter, atmometer, chloride tracer, ground water fluctuations. Micrometeorological methods include: mass transfer, energy budget, profile/gradient, and eddy correlation method. See, Section 8, Boulding (1993). **Laboratory:** Chloride tracer method requires laboratory analyses (Sharma, 1985). **Lookup:** Three commonly used empirical equations are the Thornthwaite, Blaney-Criddle, Jensen-Haise equations. The Penman-Monteith equation is a commonly used physically-based equation. Most equations require measurement or estimation of one or more meteorologic parameters. See, Boulding (1993), Sections 8.4.1 and 8.4.2.

e. Surface runoff is the amount of water leaving a site when precipitation exceeds infiltration. May be required for water budget modeling (or may be calculated from other parameters).

Field: Field observations using SCS soil runoff classes allows qualitative estimation (Tables 2-1, 2-2, 2-3 this Guide). **Laboratory:** — . **Lookup:** Can be calculated from precipitation, soil moisture and infiltration data. SCS soil series interpretation sheets provide hydrologic soil group for surface runoff calculations using the SCS curve runoff method (SCS, 1975).

2. Other Climate Parameters

a. Air temperature affects evaporation rates of water and volatile contaminants. Temperature as an influence on volatilization is important in modeling migration of volatile contaminants. May also influence sorption. See Section 14.b in this table.

Field: Use of manual (liquid-in-glass, deformation, bi-metallic, Bourdon tube) or electric (thermocouple, electrical-resistance, thermistor) thermometers (Section 8.2.1, Boulding, 1993). **Laboratory:** — . **Lookup:** Historical data from National Weather Service or other source (Hatch, 1988).

C-13

Table C-1. (Continued)

Site Parameter/Significance	Methods/References

2b. Wind speed affects rate of evaporation to atmosphere of volatile contaminants at or near the soil surface. **Wind direction** indicates highest risk locations for exposure to air-borne dust and volatile contaminants.

Field: Cup, propeller or pressure anemometers are the most commonly used types (Sections 8.2.3 and 8.2.4, Boulding, 1993). **Laboratory:** — . **Lookup:** Historical data from National Weather Service or other source (Hatch, 1988).

c. Humidity (moisture content of the air) affects the rate of evapotranspiration (ET) from the surface water and soils. High humidity = lower ET, low humidity = higher ET. Required for profile, eddy correlation and mass transfer methods for determining evapotranspiration.

Field: Measured using psychrometers (sling—ASTM E337-84, aspirated, thermocouple) or hygrometers (mechanical, dew-point/frost point—ASTM D4030-83, electric, diffusion, absorption spectra). See, Sections 8.1.3 and 8.1.4 (Boulding, 1993). **Laboratory:** — ; **Lookup:** Historical data from National Weather Service or other source (Hatch, 1988).

d. Insolation, described in units of energy flux affects evaporation rates, temperature of surface soils (and thus evaporation), and atmospheric motion. Sometimes needed for water balance calculations (Schroeder, 1983).

Field: Measured using pyranometer (thermopile, photovoltaic, bimetallic, etc.)—ASTM E824, pyrradiometer or pyrheliometer (ASTM E816-84). See, Sections 8.2.6 and 8.2.7, Boulding (1993). **Laboratory:** — . **Lookup:** Historical data from National Weather Service or other source (Hatch, 1988). Schroeder et al. (1983) include insolation data for 102 cities.

3. Geomorphology

a. Slope (%). Key input, along with slope length, for various models for estimating transport of sediment and chemicals (Mausbach and Nielsen, 1991). Important component in the design of caps to minimize erosion. Steep slopes increase erosion and lateral surface movement of contaminants; flat to nearly level slope increase infiltration and vertical movement of contaminants.

Field: Section 2.2, this guide. **Laboratory:** — . **Lookup:** Can be measured from USGS topographic maps or site topographic surveys.

Table C-1. (Continued)

Site Parameter/Significance   Methods/References

**3b. Slope form/landscape
position**. Slope shape (convex,
concave), complexity and landscape
position strongly affect the infiltration
and runoff of precipitation.

Field: Observation of slope form and
shape may help in location of areas where
contaminants have been concentrated by
surface erosion. **Laboratory**: — .
Lookup: The surface hydrology of hill
slope systems is generalized by the SCS
curve number and hydrologic soil groups
(see 1.e above).

4. Erodibility is important in the
analysis of release rate estimates
(mass per unit time) and exposure
assessment (U.S. EPA, 1988).
Release rates are necessary inputs to
environmental fate analysis. Presence
or use of soil residue/litter helps
reduce erosion and minimize off site
migration.

Field: Field measurement of slope, field
length, cover type. **Laboratory**: — .
Lookup: Estimated using standard
equations and graphs (Isrealsen et al.,
1980). Soil erodibility K factors can be
taken for SCS soil interpretation records if
soil series is known or estimated from
texture, organic matter, structure and
permeability (Figure 2-1, this guide).

**a. Water erosion (USLE or
RUSLE)**. Release rates are necessary
inputs to environmental fate analysis.
Source of concern for surface water
contamination at uncontrolled
hazardous waste sites. Calculations
useful for selection of slope, field
length and cover type to control off-
site movement of soil.

Field: Measurement/survey of slope (in. ft.
rise/ft. run or %), length of field and
vegetation cover. **Laboratory**: — ;
Lookup: The universal soil loss equation
(USLE) (Wischmeier and Smith, 1978);
revised version (Renard et al. 1991) and
the new generation WEPP (Laflen et al.
(1991). See Mills et al. (1985) and U.S.
EPA (1988) for use guidance.

b. Wind erosion. An estimate of
total wind erosion (mass/area/time) is
used with data on amount of
substance of concern in air emission
to determine importance of
atmospheric transport. Calculations
can help in selection of remedial
techniques with minimal impact to
site area.

Field: Air monitoring for mass of
contaminant. Measure field length along
prevailing wind direction for SCS wind
erosion equation (WEQ). **Laboratory**: — .
Lookup: SCS wind erosion equation
(Woodruff and Sidoway, 1965; Isrealsen et
al. 1980). Cowherd et al. (1985) describe
method for rapid evaluation of particles
from a Superfund site.

Table C-1.   (Continued)

Site Parameter/Significance	Methods/References

5. Surface Hydrology

a. Surface streams receiving runoff from contaminated sites or discharge of contaminated ground-water may allow rapid off-site transport of contaminated sediment and dissolved contaminants. Remediation measures must be designed to avoid contamination of surface streams.

Field: Field observation of drainage way location, and whether they are ephemeral, intermittent or perennial. **Laboratory:** — . **Lookup:** Air photographs, topographic and other types of maps.

b. Flood plains/frequency/duration. Contaminants in flood plains tend to be more mobile due to flooding of surface waters and the movement of contaminated ground-water toward streams. Remedial measures require protection of areas being remediated from flooding, if possible, and control of ground-water flow to prevent entrance to surface streams.

Field: Engineering and hydrologic field measurements to estimate extent and frequency of flooding. **Laboratory:** — . **Lookup:** Check for availability of flood hazard boundary maps prepared for the Federal Emergency Management Agency and the Federal Insurance Administration.

c. Water bodies such as ponds and lakes are often surface expression of the water table and serve as discharge areas for contaminated ground-water. Water bodies used for waste disposal are sources of contaminants to ground-water.

Field: Map location and measure size, and variations in water levels over time. **Laboratory:** — . **Lookup:** Air photographs, topographic and other types of maps.

Table C-1. (Continued)

Site Parameter/Significance	Methods/References

6. Biota (see Table C-4, 21.a-d for soil microbiota)

a. Vegetative cover affects the amount of evapotranspiration. Used in wind and water erosion calculations and moisture balance models. Cover type selection to minimize infiltration and maximize soil stability important in design of covers for waste disposal sites.

Field: Map current vegetation types. Vogel (1987) describes procedures for mapping and sampling vegetation. USDA can aid in identification of unknown plant species. **Laboratory:** Botanical collections may assist in identifying unfamiliar species. **Lookup:** Air photographs taken at different years can be used to evaluate changes in vegetation over time. SCS and the Agricultural Stabilization Service (ASCS) or primary sources for sequential aerial photography. May be available from other federal and state agencies.

b. Macrofauna and mesofauna. Burrowing animals create channels for preferential flow of contaminants in the soil. Not used in modeling. Presence of endangered species or other protected species may affect remediation plan development.

Field: Record visual sighting of animals, evidence of habitation, tracks, feeding remains, distribution and number of burrows with reference to source and extent of contamination. **Laboratory:** — . **Lookup:** Consult area biologists on possible presence of endangered species.

Soil Parameter/Significance	Methods/References
7. Horizons. Layered systems affect model results. Restrictive layers (e.g. hard pans) can cause perched water tables and therefore, lateral contaminant movement, as well as retarding infiltration. Multiple simplified soil profile descriptions allow assessment of soil variability.	**Field:** Observation of soil pits or soil cores (Section 3.1.1, this guide). Surface and borehole geophysical methods are useful for characterizing spatial variations in horizons. Ground penetrating radar is generally best for near-surface characterization; seismic, electrical and electromagnetic methods for deeper measurements (Section 1, Boulding, 1993). Borehole geophysical methods are useful for correlating horizons between boreholes (Section 3, Boulding, 1993). **Laboratory:** Texture analysis and chemical analyses may be required to confirm or refine horizon breaks defined in the field. X-ray radiography of undisturbed cores may be useful for identifying horizons (ASTM D4452-85). **Lookup:** SCS soil series and interpretation records.
8. USDA Texture (percentages of sand, silt and clay) has a strong influence on water, air, and contaminant movement. Soils of fine texture (clays and silts) often slow the release of water and contaminants from a site.	**Field:** Estimated by feel (Section 3.1.2 this guide). Sieves are useful for separating coarse fragments. Samples should be taken to laboratory analysis to confirm or revise field estimates. **Laboratory:** ASTM D421-85, D422-63, D1140-92, and D2217-85; Gee and Bauder (1986). Presence of organic contaminants may affect hydrometer analysis (U.S. EPA, 1987). **Lookup:** SCS soil series and interpretation records.
9. Color. Color is usually a good indicator of oxidizing and reducing conditions in the subsoil. Uniform brown or reddish colors indicate oxidizing conditions; gray colors (2 chroma or less) indicate reducing conditions; low chroma-low value colors indicate high organic matter content.	**Field:** Munsell soil color charts; Section 3.1.3, Table 3-11, this guide. **Laboratory:** Same. **Lookup:** SCS soil series and interpretation records.

Table C-2. (Continued)

Soil Parameter/Significance	Methods/References

10. Porosity (pore volume). Required for modeling of water and vapor fluxes in the soil and contaminant mass transport.

Field: Field description; Section 3.1.4, Boulding (1990); macroporosity can be characterized by application of dye tracers followed by excavation. **Laboratory**: ASTM D4404-84; gas pycnometer (Danielson and Sutherland, 1986). **Lookup**: — .

11. Zones of Increased Secondary Porosity/Permeability

a. Structure. Structural arrangement and aggregation of soil particles affect permeability of soil. Well-developed structure creates paths for more rapid movement of contaminants. **Extrastructural cracks** also provide channels for movement of contaminants.

Field: USDA soil structure classes by visual observation; Section 3.1.5a, this guide. Section 3.1.5b, this guides describes tests for identification of extrastructural cracks. **Laboratory**: Same; soil cores or samples with minimal disturbance required. **Lookup**: SCS soil series descriptions identify common structure classes for major horizons.

b. Roots. Size and density of roots affect soil permeability Channels left by dead roots create paths for more rapid movement of contaminants.

Field: USDA classes by visual observation; Section 3.1.5c, this guide. **Laboratory**: Removal of soil from monoliths; sectioning of cores, microscopic analysis of thin sections. **Lookup**: — .

c. Lateral features. Lateral features such as bleached silt coatings on ped faces indicates zones of preferential water movement. Stress features such as slickensides indicate shrinking and swelling.

Field: Visual observation of USDA types with hand lens or binocular microscope; Section 3.1.5d, this guide. **Laboratory**: Same; microscopic examination of thin sections. **Lookup**: — .

d. Sedimentary features. Non-pedogenic sedimentary features such as stratification may be indicators of zones of differential lateral movement of water and contaminants.

Field: Core/pit description; Section 3.1.5e, this guide. **Laboratory**: Core description. **Lookup**: — .

Table C-2. (Continued)

Soil Parameter/Significance	Methods/References

12. Zones of Reduced Porosity/Permeability

a. Genetic horizons. Fragipans, and cementation or induration of the soil matrix by carbonates, silica and iron restrict downward movement of contaminants.

Field: Visual observation; Sections 7 and 3.1.6a, this guide. Soil pits preferable to cores. **Laboratory**: Undisturbed core observation. **Lookup**: SCS soil series descriptions identify major genetic horizons.

b. Rupture resistance (consistence). Degree of rupture resistance and cementation of soil horizons are indicators of permeability. Firm, hard and cemented zones retard downward movement of contaminants and favor lateral movement.

Field: Tactile observation; Section 3.1.6b, this guide. **Laboratory**: Undisturbed core observations. **Lookup**: Variable property that is not readily estimated from other properties.

c. Bulk density. Basic soil descriptor; required in modeling contaminant mass transport.

Field: Core method (ASTM D2937), excavation methods (sand cone—ASTM D1556-90, rubber balloon—ASTM D2167-84, sand replacement—ASTM D41914-89), nuclear methods (surface—ASTM D2922-681, > 12"—ASTM D5196-91), coated clod method (Blake and Hartge, 1986), peat bulk density (ASTM D5422-85). Section 3.1.6c, this guide. **Laboratory**: Drying and weighing of soil material usually done in laboratory; measurement of specific gravity of materials (ASTM D854-83) may be required is soil material is unusually light or dense. **Lookup**: SCS soil series interpretation records show ranges of bulk density for major horizons.

d. Restrictive layers. Field identifiable zones other than cemented zones that restrict the depth of penetration of plant roots also indicate reduced permeability.

Field: Soil pit generally required to apply SCS criteria; Section 3.1.6.d, this guide. **Laboratory**: — . **Lookup**: — .

Table C-2. (Continued)

Soil Parameter/Significance	Methods/References

13. Soil Engineering Properties/Parameters

a. ASTM (Unified) Texture. Soil texture classification system oriented toward engineering applications of soils.

Field: Section 3.1.7, this guide describes tests for estimation in the field. See, also, ASTM DD2488-90 (visual-manual procedure). **Laboratory:** Field estimation must be confirmed by laboratory analysis (ASTM D2487-92). **Lookup:** SCS soil series interpretation records; SCS (1990).

b. Atterberg limits. Defines various states of fine-grained soil material ranging from the dry to a liquid state (plastic limit and liquid limit being the most commonly measured parameters).

Field: Ribbon, plasticity, and stickiness tests (Section 3.1.7b, this guide). **Laboratory:** ASTM D4318-84 for ASTM (Unified) classification. Other systems (AASHTO, FAA) may require slightly different procedures. **Lookup:** SCS soil series interpretation records.

c. Shear strength. Affects the maximum slope at which a soil of a particular texture will remain stable.

Field: Can be estimated from field determination of consistency (Section 3.1.7c, this guide); field vane shear test (ASTM D2573-72). **Laboratory:** ASTM D3080-90 and D4648-87. **Lookup:** — .

d. Shrink-swell (COLE). Shrinking and swelling of soils with changes in moisture content reduces soil strength and may creates cracks in engineered structures.

Field: Placement in SCS shrink-swell potential classes (Section 3.1.7d, this guide) requires collection of soil cores. **Laboratory:** Core shrinkage (SCS, 1971, 1990); see, also, ASTM D427-83, D4546-90, and D4943-89. **Lookup:** SCS soil series interpretation records.

e. Corrosivity or corrosion potential is an indication of the ability of soil to dissolve or degrade metals or construction materials such as concrete.

Field: Observation of properties affecting corrosion potential (texture, water table, drainage class—see Section 3.1.7e, this guide). Measure soil pH (ASTM G51-77) and resistivity (ASTM G57-78). **Laboratory:** Chemical and physical analyses of related parameters (acidity, conductivity, sulfates, chlorides). **Lookup:** SCS soil series interpretation records. See, also, Table 3-7, this guide.

C-21

Table C-2. (Continued)

Soil Parameter/Significance	Methods/References

13f. Compaction. Soils may be compacted to increase soil strength and decrease compressibility and/or to decrease permeability. Moisture content at the time of compaction is a critical factor.

Field: Measurement of penetration resistance with pocket penetrometer (Section 3.1.6e, this guide) or a power-driven penetrometer (ASTM D3441-86). **Laboratory:** Proctor density test (ASTM D698-91), D1557-91 or D5080-90. **Lookup:** — .

g. Compressibility. A measure of a soil's susceptibility to decrease in volume when subjected to load. Highly compressive soils (MH,CH,OH) should be avoided for engineered structures.

Field: — . **Laboratory:** ASTM D2166-91, D2850-87, and D4767-88. **Lookup:** Estimated from Unified texture class (SCS, 1990).

h. Bearing capacity. Allowable bearing capacity of a soil is the maximum average load per unit area of a footing that will not produce rupture failure or excessive settling. Affects earthmoving equipment selection and design of structures associated with remedial activities.

Field: ASTM D1194-72, and D4429-84. **Laboratory:** ASTM D1883-87. **Lookup:** — .

i. Erosion resistance. Water and air erosion are discussed in Table C-1 (Item 4). Erosion resistance to flow of water in channel affects design of open water channels and emergency spillways associated with remedial action activities.

Field: Field data required for engineering design calculations should be collected. **Laboratory:** — . **Lookup:** SCS Technical Release No. 25 (Design of Open Channels) and No. 52 (A Guide for Design and Layout of Earth Emergency Spillways).

Table C-2. (Continued)

Soil Parameter/Significance	Methods/References
13j. Frost heave. Soil movement caused by frost action may damage roads, buildings and other improvements. Mainly a problem where permeable fine-grained soils can transmit water by capillary action. Can damage improperly installed monitoring wells.	**Field**: Observation of properties affecting frost heave potential (texture and soil moisture regime). **Laboratory**: — . **Lookup**: SCS soil series interpretation records. SCS (1992) contains a table for placement of soils in frost action classes based on soil moisture regime and family texture class.

Table C-3. Soil Hydrologic Parameters of Interest and Applicable Field, Laboratory, and Calculation/Lookup Methods

Soil Parameter/Significance	Methods/References

14. Soil Water State

a. Moisture Content. This is an important component in the monitoring of field conditions in QA/QC of model validation. It can also be used to identify low and high permeability zones in unsaturated zone by locating regions of low versus high moisture content.

Field: Neutron probe (D3017-88) and carbide method (D4944-89); other: dual-gamma probe, dielectric probe, time domain reflectometry, electro-optical sensors, four-electrode resistivity, electromagnetic induction (Boulding, 1993, Section 6.2; Gardner, 1986). **Laboratory:** Gravimetric methods (may also be used in field with if suitable facilities are available): standard oven dry (D2216-90), microwave oven (D4643-87), direct heating (D4959-89). **Lookup:** Estimated from water potential measurements using specific retention curve (14c below).

b. Water Potential. The positive (saturated soil, see 10 below) or negative (unsaturated soil) hydraulic head in soil. Matric potential, or suction, (negative head) can be used to estimate moisture content using specific retention curves (9.c below), and is an important determinant in unsaturated hydraulic conductivity (water movement becomes slower as suction increases).

Field: See 15 below for measurement of positive pressure head. Porous cup tensiometers are the most commonly used method for measuring matric potential (ASTM D3404-91; Cassell and Klute, 1986). Other methods include thermocouple psychrometers (Rawlins and Campbell, 1986), resistance sensors, electrothermal methods, electro-optical sensors (Campbell and Gee, 1986). See, also, Section 6.1 in Boulding (1993). Methods for measuring moisture content (14a above) can also be calibrated for matric potential. **Laboratory:** Same as above, plus filter paper method (ASTM D5298-92). **Lookup:** Estimated from moisture measurements using specific retention curve (14c below).

Table C-3. (Continued)

Soil Parameter/Significance	Methods/References

14c. Specific Retention. Relationship of soil moisture content at different matric potentials. Used in transient modeling of water movement in soils, or where steady-state water movement cannot be assumed.

Field: Instantaneous profile, draining profile, tension infiltrometer, sprinkler/dripper methods (Boulding, 1993, Section 6.3.1; Bruce and Luxmore, 1986). **Laboratory:** Coarse- and medium-textured soils (ASTM D2325-68); fine-textured soils (ASTM D3152-72). See, also Klute (1986). **Lookup:** Estimated from various soil properties (Boulding, 1993, Section 6.3.1).

d. Available water capacity (AWC) The amount of plant- available moisture, usually expressed as inches of water per inch of soil depth, is commonly defined as the amount of water held between field capacity and the wilting point. AWC is important in developing water budgets, designing drainage systems and predicting plant yields. AWC may be a consideration in selection of soil materials for vegetation establishment.

Field: Collection of soil samples just after the soil has drained following a period of rain and humid weather, after a spring thaw, or after heavy irrigation (see Section I.4.1 in SCS, 1971). **Laboratory:** Measurement of soil moisture at field capacity and 15-bar moisture content, and bulk density (SCS, 1992). **Lookup:** SCS soil series interpretation records provide ranges of available water capacity for individual soil horizons.

e. Soil moisture regime Soil moisture regimes provide an indication of the seasonal availability of water for plant growth. SCS defines 4 moisture regimes ranging for aquic (wet the year-round) to aridic/torric where there is a moisture deficit year around. Whether a soil moisture regime is leaching (aquic, udic, xeric) or non-leaching (aridic/torric) may affect the design and implementation of remedial methods.

Field: — . **Laboratory:** — ; **Lookup:** Usually determined by using climatic and soil data to develop a monthly soil water balance. SCS taxonomic classification will indicate the moisture regime of a soil series.

Table C-3. (Continued)

Soil Parameter/Significance	Methods/References
15. Internal Free Water (Saturated Zone). Location of ground water in producing, perched or confined system affects model selection and considerations for contaminant transport.	**Field:** Measurement of water levels in ground-water monitoring wells or piezometers (Reeve, 1986): D4750-87 (steel tape, electric probe). Other methods include: air line, pressure transducers, acoustic probes, ultrasonic probes and floats (Boulding, 1993, Section 4.1). **Laboratory:** — . **Lookup:** SCS soil series interpretation records provide estimates of depth, thickness, and duration of shallow/perched water tables.

16. Permeability/Hydraulic Conductivity

a. Saturated hydraulic conductivity (Ksat). The saturated hydraulic conductivity of unsaturated soils is used in steady-state and non-steady-state modeling of water movement in soils. Ksat below the water table affects the rate of movement of contaminants in ground water	**Field:** ASTM D5156-90 provides general guidance. Above water table: ring infiltrometers (D3385-88 and D5093-90); cylinder infiltrometer, constant-head borehole infiltration, Guelph permeameter, air-entry permeameter, double tube method, infiltration gradient method, in situ monoliths Boutwell method, velocity permeameter, percolation test (Boulding, 1993, Section 7.3). Shallow water table: auger hole, piezometer and multiple holes methods (Amoozegar and Warrick, 1986; Boulding, 1993, Section 4.3). **Laboratory:** Constant- and falling-head methods (Klute and Dirksen, 1986); granular soils (D2434-68), fine textured soils (D5083-90). **Lookup:** SCS soil series interpretation records. SCS (1990) provides ranges for Unified soil texture classes. Can be estimated from various soil properties (Table 3-9, this guide).

Table C-3. (Continued)

Soil Parameter/Significance	Methods/References

16b. Unsaturated hydraulic conductivity is used in transient modeling of water movement in soils, or where steady-state water movement cannot be assumed.

Field: Commonly used methods are instantaneous profile and crust test (D5126-90). Also, draining profile methods, tension infiltrometer, sprinkler/dripper methods and entrapped air method (Boulding, 1993, Section 7.2, Green et al., 1986). **Laboratory**: Various methods using cores with steady-state head and steady-state controlled flux the most common (Klute and Dirksen, 1986). **Lookup**: Parameter identification methods and numerous physical and empirical equations and relationships can be used to estimate hydraulic conductivity functions (Mualem, 1986; Boulding, 1993, Sections 7.2.7 and 7.2.8).

17. Contaminant Transport in Soil Water

a. Water velocity Estimation of the speed at which a contaminant may move through the soil is a major objective of modeling. Model results may be useful in the selection of remedial methods. Monitoring of velocity may be required during the design and operation of remediation.

Field: Direct measurement of velocity using flow meters or tracers (Everett et al. 1982; Wilson, 1982). **Laboratory**: — ; **Lookup**: Calculation using flux values measured in field, or long-term infiltration data (Everett et al. 1982; Wilson, 1982).

b. Water/solute flux Estimation of the amount of contaminant transported into and through the soil is a major objective of modeling. Model results may be useful in the selection of remedial methods. Monitoring of flux is essential during the design and operation of remediation.

Field: Measurement of parameters required for flux calculations (infiltration, hydraulic conductivity, water content and potential changes (Wagenet, 1986; Boulding, 1993—Section 7.5). **Laboratory**: Measurement of flux through undisturbed soil cores. **Lookup**: Various methods involving water budget and soil moisture accounting, and other physical and empirical equations (Boulding, 1993—Section 7.5.6; Everett et al., 1982; Wilson, 1982).

Table C-3. (Continued)

Soil Parameter/Significance   Methods/References

17c. Dispersivity. Parameter reflecting differential rates of contaminant movement due to soil heterogeneity and diffusion processes. Results in more rapid movement of contaminants than modeling of simple advective transport.

Field: Any method used to identify zones of increased permeability (see Section 5, Table 1). Tracer tests can be used in the saturated and unsaturated zone, borehole flowmeters in the saturated zone. **Laboratory**: Use of column and flow-through tests (Van Genuchten and Wierenga, 1986). **Lookup**: Where no data are available, factors for hydrodynamic dispersion may be estimated for use in transport modeling.

18. Volatilization (see also contaminant volatilization parameters, Table C-5)

a. Air permeability at different water contents. Important consideration in the evaluation of contaminant transport through the gas phase. Air permeability is required for evaluation of site characteristics for the implementation of remediation methods where air flow is used to facilitate the transport of volatile contaminants from unsaturated soil.

Field: Field observations of porosity and soil structure and porosity serve as indicators of air permeability (Sections 3.1.4 and 3.1.5a, this guide). Monitoring os subsurface pressure response when a vacuum is applied to a borehole (Boulding, 1993—Section 9.4.5). **Laboratory**: ASTM D4525-90; various steady-state and unsteady-state methods (Corey, 1986). **Lookup**: Estimation methods similar to those for unsaturated hydraulic conductivity are available (16b above).

b. Gas diffusivity. The rate of movement of gases in the soil is described by Fick's law. Affects the rate at which volatile contaminants move through the unsaturated zone. Important factor where air stripping remediation method is used.

Field: Chamber method (Rolston et al., 1991); Boulding (1993—Section 9.4.5). **Laboratory**: Placement of field collected soil samples in diffusion chamber (Rolston, 1986b). **Lookup**: — .

Table C-3. (Continued)

Soil Parameter/Significance	Methods/References

18c. Gas flux. Estimation of the flow of volatile organic contaminants is a major objective of modeling. Methane gas flux can be used as an indicator of biodegradation of organic contaminants.

Flow: Measurement of concentrations in situ or by removal of small gas samples at depth of interest; closed-chamber and flow-through chamber methods (Rolston, 1986a). See, also, Section 9.5.2 (Boulding, 1993). **Laboratory**: — ; **Lookup**: Can be calculated using Fick's law if gas concentration and gas diffusivity of soil is known.

d. Air temperature affects evaporation rates of water and volatile contaminants. Used in modeling migration of volatile contaminants. May also influence sorption. High volatilization may rule out some remedial techniques.

Field: See 2.a. **Laboratory**: — . **Lookup**: See 2a.

e. Wind speed (see 2b)

19. Soil Temperature

a. Soil temperature. Effects evaporation rates of volatile contaminants. Indicator of recharge and discharge areas of shallow aquifers. Hotter soil temperatures tend to increase chemical reactions and volatilization.

Field: Thermometry (Taylor and Jackson, 1986); Section 3.1.8, this guide for estimation of average annual temperature. See, also, Section 1.6 in Boulding (1993). **Laboratory**: — . **Lookup**: Brown et al. (1983) provide maps with temperature data.

b. Soil temperature regime. The pattern of soil temperature fluctuations in a soil, characterized by temperature distribution with respect to depth, time, and season. SCS defines 6 major regimes from pergelic (permafrost present) to hot-moist (thermic) and hot-dry (hyperthermic).

Field: Section 3.1.8, this guide. **Laboratory**: — . **Lookup**: Usually estimated from climatic data. Soil temperature regime is part of the SCS taxonomic classification of a soil series.

Table C-4. Soil Chemical and Biological Parameters of Interest and Applicable Field, Laboratory, and Calculation/Lookup Methods

Soil Parameter/Significance	Methods/References

20. Soil Chemistry

a. Organic carbon/matter. Total organic carbon (TOC)is the amount of carbon in the soil that is contained in the soil organic fraction. Total carbon includes inorganic carbonates present as solids or in solution. Organic matter (OM) includes organic carbon-based substances in the soil and is generally 1.7 to 2.0 times TOC by weight. TOC is required for modeling the transport and migration potential of contaminants. Higher amounts of OM can enhance biological degradation of certain organic compounds; also helps bind metals at the site.

Field: Color provides a rough indication of organic matter content of soils. See color ignition test (Section 3.1.3) and Test 11 in Section 3.1.7a) in this guide. **Laboratory:** Different methods are used to measure total, inorganic and organic carbon. Most involve high temperature combustion (wet or dry) and oxidation techniques (Nelson and Sommers, 1982). Due to difficulty in identifying natural versus added organic carbon, correct analysis at contaminated sites is a problem (Powell et al., 1989; Powell, 1990). Characterization of stable humic substances in OM may be desirable (Schnitzer, 1982). ASTM D2974-87 covers determination of organic matter content of peat. **Lookup:** SCS soil series interpretation record give range of organic matter content of major horizons.

b. Odor. Organic soils have a distinctive, pungent musty odor.

Field: Used in ASTM (unified) soil classification system to identify organic soils (see Test 11 in section 3.1.7a, this guide. **Laboratory:** — . **Lookup:** — .

c. Cation exchange capacity (CEC). Needed for estimation of contaminant transport potential and sorption capacity for site; important tool for evaluating strength of metal binding to solids and method of sorption.

Field: Representative samples of major horizons and strata required for laboratory analysis. **Laboratory:** Chapman (1965) describes standard methods for measurement of CEC; Rhoades (1982a) describes special procedures for arid and highly weathered tropical soils. Thomas (1982) describes methods for determination of total exchangeable cations. **Lookup:** Rough estimates possible from particle size distribution and clay mineralogy.

C-30

Table C-4. (Continued)

Soil Parameter/Significance	Methods/References

20d. Soil pH is required input to geochemical transport models and has great effect on solute concentration and sorption/ desorption of contaminants in soils. pH adjustment during remediation can greatly reduce the mobility of some species or conversely dissolve contaminants for extraction.

Field: pH meter or pH paper (ASTM D4972-89); Section 3.3.4, this guide. **Laboratory:** Using a glass electrode in an aqueous slurry (McLean, 1982); Method 8C (SCS, 1984). **Lookup:** SCS soil series interpretation records give range of soil pH for major soil horizons.

e. Soil oxygen important where primary sulfides or other reduced species are present. As for redox potential below. Controls the rate of aerobic microbial action.

Field: Platinum or membrane electrode methods (Phene, 1986) or use of field gas chromatograph (Smith and Arah, 1991). **Laboratory:** Same as field. In saturated soils, dissolved oxygen is analyzed using EPA 360.1 or 360.2--electrometric membrane electrode and titrimetric, modified Winkler methods (Kopp and McKee, 1983). **Lookup:** Calculated from pE (Stumm and Morgan, 1981) or from oxygen and soil gas diffusion rate.

f. Eh (redox potential). In saturated soil, Eh is an indicator of soil microbial assemblages present, and of redox state of heavy metals. Needed to evaluate in-situ biotreatment, soil washing, chemical treatment. Establishes need for enhanced aeration or chemical addition.

Field: Platinum electrode used on lysimeter sample (ASTM D1498-76); Section 3.3.5, this guide. **Laboratory:** Same as field. **Lookup:** Can be calculated from concentration of redox pairs of oxygen (Stumm and Morgan, 1981).

g. Redox couple ratios (waste soil system) Essential for assessing the form and mobility of heavy metals. See Eh above.

Field: Collection of soil solute or ground-water samples for laboratory analysis. **Laboratory:** Analysis of concentrations of redox pairs. Thompson et al. (1989) summarize methods for measurement of redox sensitive species (As, Cr, Fe, and Se). **Lookup:** — .

Table C-4. (Continued)

Soil Parameter/Significance	Methods/References
20h. Clay mineralogy. Clay mineralogy strongly influences cation exchange capacity, soil shrink-swell potential, soil surface area and sorption. May be used to assess ease of removal of contaminant and sorption capacity for contaminant. Affects engineering properties of soil.	**Field:** Clay mineral and mica shine tests, Section 3.3.7, this guide. **Laboratory:** Thompson et al. (1989) review 8 methods for analyzing clay mineralogy. The most common methods are X-ray diffraction (Whittig and Allerdice, 1986), and scanning electron microscopy (Goldstein et al. 1981), sometimes supplemented with chemical analyses (Jackson et al. 1986). **Lookup:** SCS soil series taxonomic classification provides a general idea of clay mineralogy.
i. Other mineralogy. The major non-clay mineral in soil is quartz and in arid areas soluble minerals such as carbonates and gypsum may be present. Iron and manganese oxide concretions often indicate variably saturated conditions.	**Field:** Hydrochloric acid test for calcium carbonate; other mineral concentrations can be identified visually (Section 3.3.8, this guide). **Laboratory:** The most common method for identification of non-clay minerals is optical microscopy (Cady et al. 1986). Nelson (1982) describe methods for analysis of carbonate (also ASTM D4373-84) and gypsum. See also methods for clay mineralogy above. **Lookup:** SCS soil series interpretation records give $CaCO_3$ and gypsum percentages, when present in soil horizons. Soils where specific non-clay minerals form a significant percentage of the soil texture are identified in a soil series taxonomic mineralogy class.
j. Salinity (electrical conductivity) is related primarily to the amount of salts in the soil that are more soluble than gypsum.	**Field:** Measurement of electrical conductivity of a saturation extract from soil, or ground-water sample (Rhoades, 1982b). See also other methods for measurement of water flux that sense dissolved solids (see 13.a). **Laboratory:** ASTM D4542-85. **Lookup:** SCS soil series interpretation records specify soil salinity classes of major soil horizons where soluble minerals are significant in the soil profile.

Table C-4. (Continued)

Soil Parameter/Significance	Methods/References

20k. Sodicity/SAR (sodium adsorption ratio) Sodium can significantly affect soil physical properties by dispersing mineral colloids resulting in a tight, impervious soil structure (see 8.i). High concentrations of neutral soluble salts mitigate the effects of sodium, unless they are leached away.

Field: The surface of sodic soils is usually discolored by the dispersed humus carried upward by capillary water. **Laboratory:** Analysis of concentrations of sodium, calcium and magnesium in a saturation extract (SCS, 1992). See 15l below. **Lookup:** SCS soil survey interpretation records give SAR ranges where this is significant in the soil.

l. Major cations The major cations in soil are calcium, magnesium, potassium and sodium. The type and number of exchangeable cations affects sorption of metals. Cations of soluble salts in arid areas will dissolve if the amount of water entering the soil exceeds evapotranspiration.

Field: Collection of samples of soil solids or ground water for laboratory analysis required. **Laboratory:** Knudsen et al. (1982) describe methods for analyzing for sodium and potassium, and Lanyon et al. (1982) for calcium and magnesium in soil extracts. **Lookup:** — .

m. Major anions The major anions of interest in soils are nitrogen-, phosphorus-, and sulfur-containing species. In arid areas chloride and fluoride may be important.

Field: Collection of samples of soil solids or ground water for laboratory analysis required. **Laboratory:** Thompson et al. (1989) review 18 recommended methods for analysis of anionic species in aqueous solution, and 21 methods for anionic species and ammonium ion in solids. **Lookup:** — .

n. Fertility potential is the ability of soil to supply nutrients necessary for plant growth. Nutrient availability may be affected by physical factors such as texture. Low fertility may reduce plant cover, and hence increase mobility of contaminants by erosion. Metal contamination may reduce fertility by causing nutrient imbalances. Important consideration when establishing vegetative cover at the completion of remediation.

Field: Sampling of soil for laboratory analysis of nutrients of interest. **Laboratory:** Analysis for nitrogen, phosphorus and potassium (SCS, 1984). **Lookup:** Sanchez et al. (1982) describe as soil fertility capability classification system.

Table C-4. (Continued)

Soil Parameter/Significance	Methods/References

21. Soil microbiota. May transform or degrade organic contaminants and aid in mobilizing inorganic constituents. No standard procedures exist for determining degradation rates or products although some models use the result of microcosm studies to determine biotic effects. Biotic remediation technologies usually modify soil conditions experimentally in the lab or in situ to optimize degradation conditions. This may include manipulating soil moisture, nutrients, pH, oxygen content, etc. or inoculating soil with organisms.

a. Enumeration. Microorganisms in soil and ground water can be characterized in terms of species or morphology, mass, and viability. The greater the species diversity, the greater the likelihood that microorganisms can adapt to changing conditions in the subsurface.

Field: Special care required for collection of soil and ground-water samples for laboratory study (Section 3.3.10, this guide). **Laboratory:** Major types of methods include cultures, microscopic chemical and radioisotope analyses (Ghiorse and Wilson, 1988; Page et al. 1982, chapters 37 to 40; Poindexter and Ledbetter, 1986). ASTM D932, D4012, D4454, D4455. **Lookup:**—.

b. Metabolism. The ways in which microbiota metabolize contaminants indicates mechanisms of biodegradation. Organic contaminants may be mineralized (broken down to harmless inorganic forms) or transformed into toxic intermediates. Methylation of metals may increase toxicity. In situ biorestoration may requires management of subsurface conditions to ensure the most efficient metabolism of contaminants.

Field: Collection of soil solids, gases and water samples for analysis of constituents that are indicators of micro-biological activity. **Laboratory:** Measurement of oxygen respiration rates (ASTM D4478). Analysis of soil gas for byproducts of metabolism (Page et al. 1982--Chapters 41 and 42) and analysis of ground-water samples for possible transformation products of original contaminant. **Lookup:** — .

Table C-4. (Continued)

Soil Parameter/Significance	Methods/References

21c. Nutrients/terminal electron acceptor. Carbon, nitrogen, phosphorus, and sulfur are all required for microbial growth with carbon usually the greatest limiting factor in the subsurface. Organic contaminants may result in an increase in microbial population with the availability of oxygen determining whether aerobic or anaerobic degradation occur. Maintenance of aerobic biodegradation of organic contaminants may require addition of oxygen to the subsurface for insitu bioremediation.

Field: Collection of subsurface samples to identify possible nutrient deficiencies for microbiota. **Laboratory:** Analysis of constituents of interest (usually anions--see 20m above). Carbon:nitrogen: phosphorus ratios are an important indicator of potential for microbial activity. **Lookup:** — .

d. Activity/kinetics. The amount of microbial activity in the subsurface affects the rate at which organic contaminants may be metabolized. Activity is affected by the toxicity of contaminants and the ability of microorganisms to adapt to geochemical conditions created by the contamination.

Field: Collection of samples for treatability studies. **Laboratory:** Treatability studies (U.S. EPA, 1989a). **Lookup:** — .

22. Soil pollution situation. Identification of type of contaminant, concentration, and vertical distribution is a basic input of any modeling effort. Knowledge of type, depth, volume and concentration of contaminants is the basic starting point for selection and design of remediation methods.

Field: Collection of soil samples at depth and locations specified in the sampling plan using appropriate sampling tools and procedures (Chapter 4 and Appendix A, this guide). Soil pore liquid samples (ASTM D4696) and soil gas monitoring for volatile contaminants (ASTM D5314-91), when appropriate. **Laboratory:** Analysis for all constituents that are suspected of possibly being present at the site. **Lookup:** Any records related to type and handling of wastes should be checked in developing the list of constituents of concern and in choosing sampling locations.

Table C-5. Contaminant Chemical and Biological Parameters and Applicable Field, Laboratory and Calculation/Lookup Methods

Chemical Parameter/Significance	Methods/References

23. Contaminant Hazardous Properties

a. Toxicity can be measured in many ways: IDHL level (immediately dangerous to life or health), TCLP limits, LC50 (median lethal concentration), LOEL (lowest observed effect level), MATC (maximum acceptable toxicant concentration), MATE guidelines (maximum acute toxicity effluent). The known or possible presence of toxic chemicals may require use of respirators and special protective gear while sampling. Toxicity and exposure assessment are required to determine soil response action levels for remediation (U.S. EPA, 1989b).

Field: Qualitative observations of toxic effects of contaminants on plants and aquatic organisms may be observable. Use of chemical field screening methods for identification of chemicals present (Boulding, 1993—Section 10). **Laboratory:** If adequate published data are not available, laboratory toxicity methods may be required. Norton et al. (1988) describe available methods. **Lookup:** See Table C-6.

b. Reactivity. Hazardous substances may be explosive, react violently or form potentially explosive mixtures or generate toxic gases or fumes when combined with water. General chemical reactivity is discussed in 25 below. Highly reactive wastes in containers at the surface or buried require special care in handling and destruction.

Field: Field test kits (Wolbach et al., 1984). **Laboratory:** — . **Lookup:** See Table C-6.

c. Corrosivity. Very acid (less than 2 pH) and very alkaline wastes (greater than pH 12.5) are defined by EPA as corrosive. pH and buffer capacity will affect neutralization methods.

Field: Field sampling of waste to measure pH using EPA Method 9045A (U.S. EPA, 1986). **Laboratory:** — . **Lookup:** See Table C-6.

Table C-5. (Continued)

Chemical Parameter/Significance	Methods/References

23d. Ignitibility. The tendency of a substance to ignite when exposed heat.

Field: ASTM D4982-89. **Laboratory:** — . **Lookup:** See Table C-6.

24. Contaminant general chemical properties. The inherent chemical characteristics of any known or suspected contaminant at the site should be compiled for modeling and geochemical fate assessment. Most of the information can be obtained from reference books. Essential basic information for selection and design of remediation methods.

Field: — ; **Laboratory:** Laboratory methods will generally not be required unless standard reference books do not contain the desired data. **Lookup:** See below for source for specific parameters.

a. Chemical class. The chemical class (acid, base, polar neutral, nonpolar neutral, inorganic) gives a broad indication of types of geochemical behavior that may result in the subsurface.

Field: — . **Laboratory:** — . **Lookup:** See Table C-6.

b. Molecular weight/structure influence chemical behavior.

Field: — . **Laboratory:** — . **Lookup:** See Table C-6. Structure-activity relationships may be used to estimate the chemical behavior of compounds.

c. Melting/boiling point of a substance determines whether it will exist in solid, liquid or gaseous phase at temperatures in the subsurface. Gaseous phases will be mobile in the vadose zone. The mobility of liquid phases in the saturated zone depends on solubility and miscibility with water.

Field: — . **Laboratory:** — . **Lookup:** See Table C-6.

Table C-5. (Continued)

Chemical Parameter/Significance	Methods/References

24d. Specific gravity/density. Nonaqueous phase liquids (NAPLs) that are lighter than water will tend to float on the ground-water surface, and those that are denser than water will tend to sink to the bottom of an aquifer. Vapor density is a consideration in soil vacuum extraction.

Field: ASTM D5057. **Laboratory:** ASTM E1109. **Lookup:** See Table C-6. Chapter 19 in Lyman et al. (1990) covers methods for estimating densities of vapor, liquids and solids.

e. Solubility/miscibility in water. Solubility affects both organic and inorganic contaminants. Miscibility is a property of napls. High solubility and high miscibility will enhance mobility in ground-water.

Field: — . **Laboratory:** ASTM E1148. **Lookup:** See Table C-6. Water solubility can be estimated from Kow (27d below), see Chapter 2 in Lyman et al. (1990).

f. Speciation. The mobility of metals is strongly influenced by the forms they may take in the subsurface: free ions, insoluble solids, metal/ligand complexes, attachment at ion exchange site, and oxidation state.

Field: — . **Laboratory:** Chemical species analysis. **Lookup:** Phases and species in aqueous systems under various geochemical conditions can be estimated using distribution-of-species geochemical models (U.S. EPA, 1990—Chapter 5).

g. Dielectric constant is a measure of the polarizability of a material in an electric field and gives information on the capacity of a substance to store electric charge.

Field: May be required for interpretation of geophysical measurements using ground-penetrating radar or other electromagnetic methods. **Laboratory:** — . **Lookup:** Akhadov (1980), Hasted (1974), Tareev (1975).

25. Contaminant chemical reactivity (aqueous solution chemistry). Knowledge of the chemical behavior of contaminants in aqueous solution is required to model transport in water and partitioning in soil. Geochemical models may be used separately or in combination with transport models. Primary inputs for models in concentration of species present.

Field: Sampling of soil solids (for later extraction of solutions) or soil solution required for laboratory analysis. **Laboratory:** Analysis for species of interest required (see 20l and m). **Lookup:** See below for source of information on contaminant behavior in relation to specific chemical processes.

Table C-5. (Continued)

Chemical Parameter/Significance	Methods/References

25a. Acid/base reactions strongly affect pH, which in turn influences precipitation-dissolution reactions.

Field: — . **Laboratory:** Controlled laboratory tests. **Lookup:** Mills et al. (1985) describe procedures for calculating the distribution of neutral and charged forms of toxic organics.

b. Oxidation/reduction. Particularly important in evaluating mobility of metals and biodegradation of organic compounds.

Field: See 20f and g. **Laboratory:** See 20g. **Lookup:** See Table C-6.

c. Complexation. Primarily a concern in evaluating speciation and transport of metals.

Field: — . **Laboratory:** Analysis for elements and structure. **Lookup:** See Table C-6.

d. Hydrolysis occurs when a compound reacts chemically with water to form a new chemical species. Hydrolysis reactions of organic contaminants may need to be considered in selection of remediation methods.

Field: — . **Laboratory:** ASTM E895; Chapter 4 in Mill et al. (1982) describes procedures for measuring hydrolysis rate constants. **Lookup:** Mills et al. (1985) describe procedures for calculating the rate of hydrolysis from rate constants. Chapter 6 in Lyman et al. (1990) describes 6 estimation methods for certain chemical classes.

e. Catalysis. If potential reactions are mediated by a catalyst (clay minerals often perform this function) the presence or absence of the catalyst should be documented. Catalysts may be used to increase the rate of chemical treatment methods.

Field: — . **Laboratory:** Laboratory methods involve measuring reactions under controlled conditions to evaluate the effect of catalysts of interest. **Lookup:** Section 2.3.5 in U.S. EPA (1990) discusses some catalytic reactions that may affect contaminants in the subsurface.

f. Precipitation/dissolution. Of major importance in evaluating partitioning between the aqueous and solid phases. Important in methods involving soil washing.

Field: — . **Laboratory:** Batch tests simulating site conditions may be required to more fully evaluate precipitation and dissolution. **Lookup:** Use of aqueous distribution-of-species codes (U.S. EPA, 1990—Chapter 5).

Table C-5. (Continued)

Chemical Parameter/Significance	Methods/References

25g. Polymerization is the formation of large molecules by the bonding together of many smaller molecules.

Field: — . **Laboratory:** Analysis for polymerization products of chemical reactions. **Lookup:** Section 2.3.6 in U.S. EPA (1990) discusses some catalytic reactions that may affect contaminants in the subsurface.

26. Contaminant Volatilization Parameters (see also 24.c).

a. Henry's Constant (Kh). A dimensionless constant related to the solubility of gas in water. Value greater than 0.01 desirable for use of soil vacuum extraction (SVE) method.

Field: — . **Laboratory:** —. **Lookup:** See Table C-6.

b. Vapor pressure. The vapor pressure of a NAPL affects volatilization potential. Important consideration in remediation using soil vapor extraction.

Field: — . **Laboratory:** ASTM E1194. **Lookup:** See Table C-6. Chapter 14 in Lyman et al. (1990) covers estimation methods.

c. Solubility. See 24e

d. Koc sorption (gaseous phase) is an empirical coefficient used to estimate sorption of gas to organic carbon in soil. High Koc will reduce effectiveness of soil vapor extraction remediation systems.

See 27a below.

27. Soil Contaminant Sorption/Retention

a. Organic carbon partition coefficient (Koc). Empirical coefficient related to sorption of a compound and the organic carbon content of soil. Important is assessing mobility of organic contaminants. Contaminant-specific parameter. Partitioning may affect remedy selection.

Field: Soil sampling for measurement of organic carbon. **Laboratory:** ASTM E1195-87. **Lookup:** Can be estimated from Kow (see 27d below) or water solubility (Mills et al., 1985; Sims et al. 1986). See also Table C-6.

Table C-5. (Continued)

Chemical Parameter/Significance	Methods/References

27b. Linear soil/water partition coefficient (Kd). General descriptor of the significance of sorption in the waste/soil system. Needed in modeling contaminant transport. Assumes sorption is independent of concentration.

Field: Field tracer tests. Soil sampling for laboratory tests. **Laboratory**: Batch and column tests (ASTM D4319-83; Roy et al., 1992; van Genuchten and Wierenga, 1986). **Lookup**: See Table C-6.

c. Nonlinear sorption constants. Sorption is typically not linear (i.e. sorption varies with concentration, in which case sorption can be estimated using Freundlich or Langmuir equations.

Field: soil sampling for laboratory sorption tests. **Laboratory**: Empirical constants for Langmuir and Freundlich sorption equations are determined using batch or column tests at different concentrations and temperatures (ASTM D4646-87; Roy et al., 1992). **Lookup**: None available. The empirical constants are highly substance and solid specific.

d. Octanol-water partition coefficient (Kow). Used in some equations for estimating Koc and water solubility for organic compounds where better data are not available.

Field: — . **Laboratory**: ASTM E1147; Chapter 1 in Lyman et al. (1990). **Lookup**: Can sometimes be estimated with solvent regression equations or estimated activity coefficients (Lyman et al., 1990).

e. Residual saturation (Ko). Important parameter in modeling the movement of nonaqueous phase liquids in the subsurface. Affects the feasibility of soil flushing.

Field: Sampling for laboratory analysis. **Laboratory**: Thermal extraction to determine amount held on soil particles. **Lookup**: — .

Table C-5. (Continued)

Chemical Parameter/Significance	Methods/References

28. Soil Contaminant Degradation

a. Half-life/rate constant is a measure of degradation or transformation by hydrolysis or biodegradation. Half-life is the time is takes for a contaminant to degrade to half its original concentration. Rate constants are empirically measured factors that describe the rate at which degradation takes place. Both are contaminant and/or site specific. Short half-lives may obviate the need for remediation.

Field: — . **Laboratory:** See 25d above and 28b below. **Lookup:** See Table C-6.

b. Biodegradability. Many organic contaminants can be metabolized by microorganisms. Affects selection of methods involving bioremediation. It may be necessary to control redox conditions to favor degrading bacteria.

Field: Sampling for laboratory tests. **Laboratory:** ASTM D1196 (anaerobic) and E1279. Chapter 9 in Lyman et al. (1990) describes typical test methods. **Lookup:** See Table C-6.

c. Degradation daughter products. When contaminants are not degraded to harmless inorganic constituents, the types and mobility of degradation products needs to be evaluated. Remediation methods must be able to handle degradation daughter products.

Field: Sampling for laboratory analysis to detect degradation products. **Laboratory:** Test procedures for hydrolysis or biodegradation with added analyses for identification of degradation products. **Lookup:** See Table C-6.

Table C-6. Index for Major References on Contaminant Chemical Properties and Fate Data

Topic	References
General Chemical Properties	ACS (annual), Budavari (1989), Dean (1992), Howard and Neal (1992), Lewis (1992a), Lide (1993), Perry and Chiltin (1973), Verschueren (1983)
Hazardous Chemicals	ACGIH (1992), Armour (1991), Government Institutes (annual), Keith (1993), Keith and Walters (1991), Lee (1993), Lewis (1990, 1991, 1992b, 1993), NIOSH (1990), Occupational Safety Health Services (1990), Patnalk (1992), Richardson and Gangolli (1993), Shafer (1993), Shineldecker (1992), U.S. Coast Guard (1985), U.S. DOT (1990), U.S. EPA (1985, 1992); Agrochemicals: Fisher (1991), James and Kidd (1992), Kidd and James (1991), Montgomery (1993), Walker and Keith (1992)
Chemical Fate Data	Callahan et al. (1979), Gherini et al. (1988, 1989), Howard (1989, 1990a, 1990b, 1992, 1993), Howard et al. (1991), Lyman et al. (1990), Mabey et al. (1982), Mackay et al. (1991, 1992), Montgomery (1991), Montgomery and Welkom (1989), Ney (1990), Rai and Zachara (1984), U.S. EPA (1990); Sorption/Partition Coefficients: Leo et al. (1971), Sablić (1988); Henry's Law Constants: Yaws et al. (1991)

Appendix C References (See Appendix E for ASTM standards)

Akhadov, Y. 1980. Dielectric Properties of Binary Solutions. Pergamon, New York, 475 pp.

American Conference of Governmental Industrial Hygienists (ACGIH). 1992. 1992-1993 Threshold Limit Values for chemical Substances and Physical Agents and Biological Exposure Indices. ACGIH, Technical Information Office, 6500 Glenway Ave., Bldg. D-7, Cincinnati, OH 45211-4438.

American Chemical Society (ACS). Annual. Chemcyclopedia: The Manual of Commercially Available Chemicals. ACS, Washington, DC.

Amoozegar, A. and A.W. Warrick. 1986. Hydraulic Conductivity of Saturated Soils: Field Methods. In: Methods of Soil Analysis, Part 1, 2nd ed., A. Klute (ed.), Agronomy Monograph No. 9, American Society of Agronomy, Madison, WI, pp. 735-770.

Armour, M.A. 1991. Hazardous Laboratory Chemicals Disposal Guide. CRC Press, Boca Raton, FL, 464 pp.

Blake, G.R. and K.H. Hartge. 1986. Bulk Density. In: Methods of Soil Analysis, Part 1, 2nd ed., A. Klute (ed.), Agronomy Monograph No. 9, American Society of Agronomy, Madison, WI, pp. 363-275.

Boulding, J.R. 1993. Subsurface Characterization and Monitoring Techniques: Vol. I, Solids and Ground Water; Vol. II, The Vadose Zone, Field Screening and Analytical Methods. EPA/625/R-93/003a&b. Available from CERI.*

Bouwer, H. 1986. Intake Rate: Cylinder Infiltrometer. In: Methods of Soil Analysis, Part 1, 2nd ed., A. Klute (ed.), Agronomy Monograph No. 9, American Society of Agronomy, Madison, WI, pp. 825-844.

Brown, K.W., G.B. Evans, Jr., and B.D. Frentrup (eds.). 1983. Hazardous Waste Land Treatment, Rev. ed. EPA 530/SW-874 (NTIS PB89-179014). Also published under the same title by Butterworth Publishers, Boston.

Bruce, R.R. and R.J. Luxmore. 1986. Water Retention: Field Methods. In: Methods of Soil Analysis, Part 1, 2nd ed., A. Klute (ed.), Agronomy Monograph No. 9, American Society of Agronomy, Madison, WI, pp. 663-686.

Budavari, S. (ed.). 1989. The Merck Index: An Encyclopedia of Chemicals, Drugs, and Biologicals, 11th ed. Merck and Co., Rahway, NJ 07065. [[Around 10,000 listings with extensive index and cross-index]

Cady, J.G., L.P. Wildung, and L.R. Drees. 1986. Petrographic Microscope Techniques. In: Methods of Soil Analysis, Part 1, 2nd ed., A. Klute (ed.), Agronomy Monograph No. 9, American Society of Agronomy, Madison, WI, pp. 185-218.

Callahan, M.A. et al. 1979. Water-Related Environmental Fate of 129 Priority Pollutants, 2 Volumes. EPA 440/4-79/029a-b (NTIS PB80-204373 and PB80-204381).

Cassell, D.K. and A. Klute. 1986. Water Potential: Tensiometry. In: Methods of Soil Analysis, Part 1, 2nd ed., A. Klute (ed.), Agronomy Monograph No. 9, American Society of Agronomy, Madison, WI, pp. 563-596.

Campbell, G.S. and G.W. Gee. 1986. Water Potential: Miscellaneous Methods. In: Methods of Soil Analysis, Part 1, 2nd ed., A. Klute (ed.), Agronomy Monograph No. 9, American Society of Agronomy, Madison, WI, pp. 619-633.

Chapman, H.D. 1965. Cation Exchange Capacity. In: Methods of Soil Analysis, (1st ed), A. Black, et al., (eds.), American Society of Agronomy, Madison, WI, pp. 891-901.

Corey, A.T. 1986. Air Permeability. In: Methods of Soil Analysis, Part 1, 2nd ed., A. Klute (ed.), Agronomy Monograph No. 9, American Society of Agronomy, Madison, WI, pp. 1121-1136.

Cowherd, C. G.E. Mulseki, P.J. Englehart, and D.A. Gillette. 1985. Rapid Assessment of Exposure to Particulate Emissions from Surface Contamination Sites. Prepared by Midwest Research Institute, Kansas City, MO. NTIS PB85-192219.

Danielson, R.E. and P.L. Sutherland. 1986. Porosity. In: Methods of Soil Analysis, Part 1, 2nd ed., A. Klute (ed.), Agronomy Monograph No. 9, American Society of Agronomy, Madison, WI, pp. 443-461.

Dean, J.A. (ed.). 1992. Lange's Handbook of Chemistry, 14th ed. McGraw-Hill, New York, 1472 pp. [Data on chemical and physical properties of elements, minerals, inorganic compounds, organic compounds, and miscellaneous tables of specific properties]

Dunne, T. and L.B. Leopold. 1978. Water in Environmental Planning. W.H. Freeman, San Francisco, 818 pp.

Everett, L.G., L.G. Wilson, and L.G. McMillion. 1982. Vadose Zone Monitoring Concepts for Hazardous Waste Sites. Ground Water 20(3):312-324.

Fisher, N. (ed.). 1991. Farm Chemicals Handbook '91. Meister Publishing Co., Willoughby, OH, 216/942-2000. [Pesticides and Fertilizers]

Gardner, W.H. 1986. Water Content. In: Methods of Soil Analysis, Part 1, 2nd ed., A. Klute (ed.), Agronomy Monograph No. 9, American Society of Agronomy, Madison, WI, pp. 493-544.

Gee, G.W. amd J.W. Bauder. 1986. Particle-Size Analysis. In: Methods of Soil Analysis, Part 1, 2nd ed., A. Klute (ed.), Agronomy Monograph No. 9, American Society of Agronomy, Madison, WI, pp. 383-411.

Gherini, S.A., K.V. Summers, R.K. Munson, and W.B. Mills. 1988. Chemical Data for Predicting the Fate of Organic Compounds in Water, Vol. 2: Database. EPRI EA-5818. Electric Power Research Institute, Palo Alto, CA. [Data relevant to predicting the release, transport, transformation and fate of more than 50 organic compounds]

Gherini, S.A., K.V. Summers, R.K. Munson, and W.B. Mills. 1989. Chemical Data for Predicting the Fate of Organic Compounds in Water, Vol. 1: Technical Basis. EPRI EA-5818. Electric Power Research Institute, Palo Alto, CA.

Ghiorse, W.C. and J.T. Wilson. 1988. Microbial Ecology of the Terrestrial Subsurface. Advances in Applied Microbiology 33:107-172.

Goldstein, J.I., D.E. Newbury, P. Echlin, D.C. Joy, C. Fiori, and E. Lifshin. 1981. Scanning Electron Microscopy and X-Ray Microanalysis. Plenum Press, New York, 673 pp.

Government Institutes, Inc. Annual. Book of Lists for Regulated Hazardous Substances, 1993 ed. Government Institutes, Inc., 4 Research Place, Suite 200, Rockville, MD, 20850; 301/921-2355, 345 pp. [Contains 70 regulatory lists of hazardous substances; updated annually]

Green, W.H. and C.A. Ampt. 1911. Studies on Soil Physics, I: Flow of Air and Water through Soils. J. Agricultural Science 4:1-24.

Green, R.E., L.R. Ahuja, and S.K. Chong. 1986. Hydraulic Conductivity, Diffusivity, and Sorptivity of Unsaturated Soils: Field Methods. In: Methods of Soil Analysis, Part 1, 2nd ed., A. Klute (ed.), Agronomy Monograph No. 9, American Society of Agronomy, Madison, WI, pp. 771-798.

Hasted, T. 1974. Aqueous Dielectrics. Chapman Hall, London.

Hatch, W.L. 1988. Selective Guide to Climatic Data Sources. Key to Meteorological Records Documentation No. 4.11. NOAA National Climate Data Center, Asheville, NC.

Howard, P.H. (ed.). 1989. Handbook of Environmental Fate and Exposure Data for Organic Chemicals: Vol. I, Large Production and Priority Pollutants. Lewis Publishers, Chelsea, MI, 600 pp.

Howard, P.H. (ed.). 1990a. Handbook of Environmental Fate and Exposure Data for Organic Chemicals: Vol. II, Solvents. Lewis Publishers, Chelsea, MI, 536 pp.

Howard, P.H. (ed.). 1990b. Handbook of Environmental Fate and Exposure Data for Organic Chemicals: Vol. III, Pesticides. Lewis Publishers, Chelsea, MI, 712 pp.

Howard, P.H. 1992. PC Environmental Fate Databases: Datalog, Chemfate, Biolog, and Biodeg. Lewis Publishers, Chelsea, MI. [Each database comes with a manual and diskettes: **Datalog** contains 180,00 records for 13,000 chemicals; **Chemfate** contains actual physical property values and rate constants for 1700 chemicals; **Biolog** contains 40,000 records on microbial toxicity and biodegradation data on about 6,000 chemicals; **Biodeg** contains data on biodegradation studies for about 700 chemicals]

Howard, P.H. (ed.). 1993. Handbook of Environmental Fate and Exposure Data for Organic Chemicals: Vol. IV, Solvents 2. Lewis Publishers, Chelsea, MI, 608 pp.

Howard, P.H. and M.W. Neal. 1992. Dictionary of Chemical Names and Synonyms. Lewis Publishers, Chelsea, MI, 2544 pp. [Basic information on more than 20,000 chemicals]

Howard, P.H., W.F. Jarvis, W.M. Meylan, and E.M. Mikalenko. 1991. Handbook of Environmental Degradation Rates. Lewis Publishers, Chelsea, MI, 700+ pp. [[Provides rate constants and half-life ranges for different media for more than 430 organic chemicals; processes include aerobic and anaerobic degradation, direct photolysis, hydrolysis and reaction with various oxidants or free radicals]

Isrealsen, C.E., C.G. Clyde, J.E. Fletcher, E.K. Isrealsen, F.W. Haws, P.E. Packer, and E.E. Farmer. 1980. Erosion Control During Highway Construction: Manual on Principles and Practices. Transportation Research Board, National Research Council, Washington, DC.

Jackson, M.L., C.H. Lim, and L.W. Zelazny. 1986. Oxides, Hydroxides, and Alumninosilcates. In: Methods of Soil Analysis, Part 1, 2nd ed., A. Klute (ed.), Agronomy Monograph No. 9, American Society of Agronomy, Madison, WI, pp. 101-150.

James, D.R. and H. Kidd. 1992. Pesticide Index, 2nd ed. Lewis Publishers/Royal Society of Chemistry, Chelsea, MI, 288 pp. [Listing of about 800 active ingredients and 25,000 trades of pesticides containing the ingredients]

Keith, L.H. (ed.). 1993. IRIS: EPA's Chemical Information Database. Lewis Publishers, Chelsea, MI. [Manual and annual subscription product updated on a quarterly basis; information on acute hazard information and physical and chemical properties on around 500 regulated and unregulated hazardous substances]

Keith, L. and D.B. Walters. 1991. The National Toxicology Program's Chemical Database: Vol. 1, Chemical Names and Synonyms; Vol. 2, Physical and Chemical Properties; Vol. 3 Standards and Regulations; Vol. 4, Medical Hazards and Symptoms of Exposure; Vol. 5, Medical First Aid; Vol. 6, Personal Protective Equipment; Vol. 7, Hazardous Properties and Uses; Vol. 8, Shipping Classifications and Regulations. Lewis Publishers, Chelsea, MI. [Covers 2,270 chemicals; each volume available separately on diskette or hard copy]

Kidd, H. and D.R. James (eds.). 1991. The Agrochemicals Handbook, 3rd ed. Lewis Publishers/Royal Society of Chemistry, Chelsea, MI, 1500 pp.

Klute, A. 1986. Water Retention: Laboratory Methods. In: Methods of Soil Analysis, Part 1, 2nd ed., A. Klute (ed.), Agronomy Monograph No. 9, American Society of Agronomy, Madison, WI, pp. 635-662.

Klute, A. and C. Dirksen. 1986. Hydraulic Conductivity and Diffusivity: Laboratory Methods. In: Methods of Soil Analysis, Part 1, 2nd ed., A. Klute (ed.), Agronomy Monograph No. 9, American Society of Agronomy, Madison, WI, pp. 687-734.

Kopp, J.F., and G.D. McKee. 1983. Methods for Chemical Analysis of Water and Wastes. EPA600/4-74-020 (NTIS PB84-128677). [Supersedes report with the same title dated 1979].

Knudsen, D., G.A. Peterson, and P.F. Pratt. 1982. Lithium, Sodium, and Potassium. In: Methods of Soil Analysis, Part 2, 2nd ed., A.L. Page et al. (eds.), Agronomy Monograph No. 9, American Society of Agronomy, Madison, WI, pp. 225-246.

Laflen, J.M., L.J. Lane, and G.R. Foster. 1991. WEPP: A New Generation of Erosion Prediction Technology. J. Soil and Water Conservation 46(1):34-38.

Lanyon, L.E. and W.R. Heald. 1982. Magnesium, Calcium, Strontium, and Barium. In: Methods of Soil Analysis, Part 2, 2nd ed., A.L. Page et al. (eds.), Agronomy Monograph No. 9, American Society of Agronomy, Madison, WI, pp. 247-262.

Lee, C.C. 1993. Environmental Law Index to Chemicals. Government Institutes, Inc., Rockville, MD, 250 pp. [Keys to which regulations govern a particular chemical]

Leo, A., C. Hansch, and D. Elkins. 1971. Partition Coefficients and Their Uses. Chemical Reviews 71(6):525-616. [First major literature review on partition coefficients and their uses; compilation of coefficients from more than 500 references]

Lewis, R.J., Sr. 1990. Carcinogenically Active Chemicals: A Reference Guide. Van Nostrand Reinhold, New York, 1184 pp. [Information on more than 3,400 chemicals]

Lewis, R.J., Sr. 1991. Reproductively Active Chemicals. Van Nostrand Reinhold, New York, 1184 pp. [Information on about 3,500 chemicals]

Lewis, Sr., R.J. 1992a. Hawley's Condensed Chemical Dictionary, 12th ed. Van Nostrand Reinhold, New York, 1288 pp. [More than 19,000 entries on chemicals, reactions and processes, state of matter, compounds. N.I. Sax and R.J. Lewis were authors of 11th edition, published in 1987]

Lewis, Sr., R.J. 1992b. Sax's Dangerous Properties of Industrial
   Materials, 8th ed (3 Volumes). Van Nostrand Reinhold, NY,
   4300 pp. [Contains some 20,000 chemical entries covering
   physical and carcinogenic properties, clinical aspects, exposure
   standards, and regulations. N.I. Sax and R.J. Lewis were authors
   of 7th edition, published in 1989. Earlier editions: 1963 (2nd),
   1968 (3rd), 1975 (4th), 1976 (5th), 1984 (6th)]

Lewis, Sr., R.J. 1993. Hazardous Chemicals Desk Reference, 3rd ed.
   Van Nostrand Reinhold, New York, 1752 pp. [Covers more than
   6,000 of the most hazardous chemicals; each entry provides the
   chemical's hazard rating, a toxic and hazard review paragraph,
   CAS, NIOSH and DOT numbers, description of physical
   properties, synonyms, and current standards for exposure limits.
   Lewis was author of 2nd edition, published in 1990; N.I. Sax and
   R.J. Lewis were authors of 1st edition, published in 1987]

Lide, D.R. 1993. CRC Handbook of Chemistry and Physics, 74th ed.
   CRC Press, Boca Raton, Fl, 2472 pp. [New edition published
   annually]

Lyman, W.J., W.F. Reehl, and D.H. Rosenblatt (eds.). 1990. Handbook
   of Chemical Property Estimation Methods: Environmental
   Behavior of Organic Compounds, 2nd ed. American Chemical
   Society, Washington, DC, 960 pp. [First edition published by
   McGraw-Hill in 1982].

Mabey, W.R., et al. 1982. Aquatic Fate Process Data for Organic Priority
   Pollutants. EPA 440/4-81-014 (NTIS PB87-169090).

Mackay, D., W.Y. Shiu, and K-C. Ma. 1991. Illustrated Handbook of
   Physical-Chemical Properties and Environmental Fate for Organic
   Chemicals, Vol. I: Monoaromatic Hydrocarbons, Chlorobenzenes,
   and PCBs. Lewis Publishers, Chelsea, MI, 704 pp.

Mackay, D., W.Y. Shiu, and K-C. Ma. 1992. Illustrated Handbook of
   Physical-Chemical Properties and Environmental Fate for Organic
   Chemicals, Vol. II: Polynuclear Aromatic Hydrocarbons,
   Polychlorinated Dioxins, and Dibenzofurans. Lewis Publishers,
   Chelsea, MI, 608 pp.

Mausbach, M.J. and R.D. Nielsen. 1991. Some Concepts Concerning Soil
   Site Assessment for Water Quality. Soil Survey Horizons
   32(1):18-25.

McLean, E.O. 1982. Soil pH and Lime Requirement. In: Methods of Soil Analysis, Part 2, 2nd ed., A.L. Page et al. (eds.), Agronomy Monograph No. 9, American Society of Agronomy, Madison, WI, pp. 199-224.

Mill, T., W.R. Mabey, D.C. Bomberger, T.-W Chou, D.G. Hendry, and J.H. Smith. 1982. Laboratory Protocols for Evaluating the Fate of Organic Chemicals in Air and Water. EPA/600/3-83-022 (NTIS PB83-150888).

Mills, W.B., et al. 1985. Water Quality Assessment: A Screening Procedure for Toxic and Conventional Pollutants in Surface and Ground Water--Part I (Revised 1985). EPA/600/6-85/002a [Updates 1982 version EPA/600/6-82/004].

Montgomery, J.H. 1991. Ground Water Chemical Desk Reference, Vol. 2. Lewis Publishers, Chelsea, MI. [Data on 267 additional compounds not included in Montgomery and Welkom (1989)]

Montgomery, J.H. 1993. Agrochemicals Desk Reference: Environmental Data. Lewis Publishers, Chelsea, MI, 672 pp. [Physical/chemical data on 200 compounds including pesticide, herbicides and fungicides, partition coefficients, transformation products, etc.]

Montgomery, J.H. and L.M. Welkom. 1989. Ground Water Chemicals Desk Reference. Lewis Publishers, Chelsea, MI. [Data on 137 organic compounds commonly found in ground water and the unsaturated zone, include: appearance, odor, boiling point, dissociation constant, Henry's law constant, log Koc, Log Kow, melting point, solubility in water and organics, specific density, transformation products, vapor pressure, fire hazard data (lower and upper explosive limits), and health hazards (IDLH, PEL). See, also, Montgomery (1991)]

Mualem, Y. 1986. Hydraulic Conductivity of Unsaturated Soils: Prediction and Formulas. In: Methods of Soil Analysis, Part 1, 2nd ed., A. Klute (ed.), Agronomy Monograph No. 9, American Society of Agronomy, Madison, WI, pp. 799-823

Musgrave, G.W. and H.N. Holtan. 1964. Infiltration. In: Handbook of Applied Hydrology, V.T. Chow (ed.), McGraw-Hill, New York, pp. 12-1 to 12-30.

National Institute for Occupational Safety and Health (NIOSH). 1985. NIOSH Pocket Guide to Chemical Hazards. DHHS (NIOSH) Publication No. 85-114, 249 pp. [Summarizes information from the three-volume **NIOSH/OSHA Occupational Health Guidelines for Chemical Hazards**; data are presented in tables, and the sourceincludes chemical names and synonyms, permissible exposure limits, chemical and physical properties and other toxicological information]

National Weather Service. 1972. Observing Handbook No. 2. Data Acquisition Division, Office of Meteorological Operations, Silver Spring, MD.

Nelson, R.E. 1982. Carbonate and Gypsum. In: Methods of Soil Analysis, Part 2, 2nd ed., A.L. Page et al. (eds.), Agronomy Monograph No. 9, American Society of Agronomy, Madison, WI, pp. 181-197.

Nelson, D.W. and L.E. Sommers. 1982. Total Carbon, Organic Carbon, and Organic Matter. In: Methods of Soil Analysis, Part 2, 2nd ed., A.L. Page et al. (eds.), Agronomy Monograph No. 9, American Society of Agronomy, Madison, WI, pp. 539-579.

Ney, Jr., R.E. 1990. Where Did That Chemical Go? A Practical Guide to Chemical Fate and Transport in the Environment. Van Nostrand Reinhold, New York, 200 pp. [Information on more than 100 organic and inorganic chemicals]

Norton, S., M. McVey, J. Colt, J. Durda, and R. Hegner. 1988. Review of Ecological Risk Assessment Methods. EPA/230/10-88/041.

Occupational Safety Health Services. 1990. PESTLINE: Material Safety Data Sheets for Pesticides and Related Chemicals, 2 Vols. Van Nostrand Reinhold, New York, 2100 pp. [Information on about 1,200 pesticides]

Page, A.L., R.H. Miller, and D.R. Keeney. 1982. Methods of Soil Analysis, Part 2: Chemical and Biological Properties, 2nd ed. Agronomy Monograph No. 9, American Society of Agronomy, Madison, WI, 1159 pp.

Patnalk, P. 1992. A Comprehensive Guide to the Hazardous Properties of Chemical Substances. Van Nostrand Reinhold, New York, 800 pp. [Information on the 1,000 most commonly encountered hazardous chemicals]

Perry, H.P. and C.H Chiltin (eds.). 1973. Chemical Engineers Handbook. McGraw-Hill, New York.

Peterson, A.E. and G.D. Bubenzer. 1986. Intake Rate: Sprinkler Infiltrometer. In: Methods of Soil Analysis, Part 1, 2nd ed., A. Klute (ed.), Agronomy Monograph No. 9, American Society of Agronomy, Madison, WI, pp. 845-870.

Phene, C.J. 1986. Oxygen Electrode Measurement. In: Methods of Soil Analysis, Part 1, 2nd ed., A. Klute (ed.), Agronomy Monograph No. 9, American Society of Agronomy, Madison, WI, pp. 1137-1159.

Philip, J.R. 1957. The Theory of Infiltration, I: The Infiltration Equation and its Solution. J. Soil Science 83:345-357.

Poindexter, J.S. and E.R. Leadbetter (eds.). 1986. Bacteria in Nature, Vol. 2: Methods and Special Applications in Bacterial Ecology. Plenum, New York, 385 pp.

Powell, R.M. 1990. Total Organic Carbon Determinations in Natural and Contaminated Aquifer Materials, Relevance and Measurement. In: Proc. Fourth Nat. Outdoor Action Conf. on Aquifer Restoration, Ground Water Monitoring and Geophysical Methods, National Water Well Association, Dublin, OH, pp. 1245-1258.

Powell, R.M., B.E. Bledsoe, R.L. Johnson, and G.P. Curtis. 1989. Interlaboratory Methods Comparison for the Total Organic Carbon Analysis of Aquifer Materials. Environ. Sci. Technol. 23:1246-1249.

Rai, D. and J.M. Zachara. 1984. Chemical Attenuation Rates, Coefficients and Constants in Leachate Migration. Vol. 1: A Critical Review. EPRI EA-3356. Electric Power Research Institute, Palo Alto, CA. [Data on 21 elements related to leachate migration: Al, Sb, As, Ba, Be, B, Cd, Cr, Cu, F, Fe, Pb, Mn, Hg, Mo, Ni, Se, Na, S, V, and Zn; see Rai et al. (1984) for annotated bibliography]

Rawlins, S.L. and G.S. Campbell. 1986. Water Potential: Thermocouple Psychometry. In: Methods of Soil Analysis, Part 1, 2nd ed., A. Klute (ed.), Agronomy Monograph No. 9, American Society of Agronomy, Madison, WI, pp. 597-618.

Reeve, R.C. 1986. Water Potential: Piezometry. In: Methods of Soil Analysis, Part 1, 2nd ed., A. Klute (ed.), Agronomy Monograph No. 9, American Society of Agronomy, Madison, WI, pp. 545-561.

Renard, K.G., G.R. Foster, G.A. Weesies, and J.P. Porter. 1991. RUSLE: Revised Universal Soil Loss Equation. J. Soil and Water Conservation 46(1):30-33.

Rhoades, J.D. 1982a. Cation Exchange Capacity. In: Methods of Soil Analysis, Part 2, 2nd ed., A.L. Page et al. (eds.), Agronomy Monograph No. 9, American Society of Agronomy, Madison, WI, pp. 149-157.

Rhoades, J.D. 1982b. Soluble Salts. In: Methods of Soil Analysis, Part 2, 2nd ed., A.L. Page et al. (eds.), Agronomy Monograph No. 9, American Society of Agronomy, Madison, WI, pp. 167-179.

Richardson, M.I. and S. Gangolli. 1993. Dictionary of Substances and their Effects, Vol. I, A-B. Lewis Publishers/Royal Society of Chemistry, Chelsea, MI, 968 pp. [First of 7 volumes with physico-chemical, ecotoxicity, and mammalian toxicity on nearly 6,000 chemical substances. Complete set to be available by 1995]

Rolston, D.E. 1986a. Gas Flux. In: Methods of Soil Analysis, Part 1, 2nd ed., A. Klute (ed.), Agronomy Monograph No. 9, American Society of Agronomy, Madison, WI, pp. 1103-1119.

Rolston, D.E. 1986b. Gas Diffusivity. In: Methods of Soil Analysis, Part 1, 2nd ed., A. Klute (ed.), Agronomy Monograph No. 9, American Society of Agronomy, Madison, WI, pp. 1089-1102.

Roy, W.R., I.G. Krapac, S.F.J. Chou, and R.A. Griffin. 1992. Batch-Type Adsorption Procedures for Estimating Soil Adsorption of Chemicals. Technical Resource Document EPA/530-SW-87/006-F (NTIS PB92-146190).

Sablić, A. 1988. On the Prediction of Soil Sorption Coefficients of Organic Pollutants by Molecular Topology. Environ. Sci. Technol. 21(4):358-366. [Sorption coefficient data for 72 nonpolar and 159 polar and ionic organic compounds]

Sanchez, P.A., W. Cuoto, and S.W. Buol. 1982. The Fertility Capability Soil Classification System: Interpretation, Applicability and Modification. Geoderma 27:283-309.

Sax and Lewis (1987)—see Lewis (1992a)

Sax and Lewis (1989)—see Lewis (1992b)

Schnitzer, M. 1982. Organic Matter Characterization. In: Methods of Soil Analysis, Part 2, 2nd ed., A.L. Page et al. (eds.), Agronomy Monograph No. 9, American Society of Agronomy, Madison, WI, pp. 581-594.

Schroeder, P.R., A.C. Gibson, and M.P. Smolen. 1983. The Hydrologic Evaluation of Landfill Performance (HELP) Model. EPA/DF-85-001.

Shafer, D. (ed.). 1993. The Book of Chemical Lists. Business & Legal Reports, Inc., Madison, CT, 800/727-5257. [Two loose-leaf volumes: Section I (Master Chemical Cross-Reference), Section II (Environmental Planning and Reporting), Section III (Health and Safety Guidelines), Section IV (State chemical lists); updated annually, supplements available for earlier editions]

Sharma, M.L. 1985. Estimating Evapotranspiration. In: Advances in Irrigation, 3. Academic Press, New York.

Shineldecker, C.L. 1992. Handbook of Environmental Contaminants. Lewis Publishers, Chelsea, MI, 371 pp. [Key to contaminants that are likely to be associated with specific types of facilities, processes, and products]

Sims, R.C., et al. 1986. Contaminated Surface Soils In-Place Treatment Techniques. Pollution Technology Review No. 132. Noyes Publications, Park, Ridge, NJ, 536 pp.

Soil Conservation Service (SCS). 1971. Handbook of Soil Survey Investigations Procedures.

Soil Conservation Service (SCS). 1975. Surface Runoff. In: Engineering Field Manual for Conservation Practices, SCS, Washington, DC, Chapter 2.

Soil Conservation Service (SCS). 1984. Procedures for Collecting Soil Samples and Methods of Analysis for Soil Survey. Soil Survey Investigations Report No. 1. U.S. Government Printing Office.

Soil Conservation Service (SCS). 1990. Elementary Soil Engineering. In: Engineering Field Manual for Conservation Practices, SCS, Washington, DC, Chapter 4.

Soil Conservation Service (SCS). 1992. Draft National Soil Survey Interpretations Handbook. [Available for inspection at SCS State, Area, and possibly County Offices]

Smith, K.A. and J.R.M. Arah. 1991. Gas Chromatographic Analysis of Soil Atmosphere. In: Soil Analysis: Modern Instrumental Techniques, 2nd ed., K.A. Smith (ed.), Marcel Dekker, New York, pp. 505-546. [First edition published 1983].

Stumm, W. and J.J. Morgan. 1981. Aquatic Chemistry, 2nd ed. Wiley-Interscience, NY.

Tareev, B. 1975. Physics of Dielectric Materials. Mir, Moscow.

Taylor, S.A. and R.D. Jackson. 1986. Temperature. In: Methods of Soil Analysis, Part 1, 2nd ed., A. Klute (ed.), Agronomy Monograph No. 9, American Society of Agronomy, Madison, WI, pp. 927-940.

Thomas, G.W. 1982. Exchangeable Cations. In: Methods of Soil Analysis, Part 2, 2nd ed., A.L. Page et al. (eds.), Agronomy Monograph No. 9, American Society of Agronomy, Madison, WI, pp. 159-165.

Thompson et al. 1989. Techniques to Develop Data for Hydrogeochemical Models. EPRI EN-6637. Electric Power Research Institute, Palo Alto, CA.

U.S. Army Corps of Engineers. 1980. Simplified Field Procedures for Determining Vertical Moisture Flow Rates in Medium to Fine Textured Soils. Engineer Technical Letter, 21 pp. [Infiltration test basins]

U.S. Coast Guard. 1985. CHRIS: Chemical Hazard Response Information System: Vol. 1, Condense Guide to Chemical Hazards (CG-446-1); Vol. 2, Hazardous Substance Data Manual (CG-446-2—3 binders, GPO Stock No. 050-012-00147-2); Vol. 3, Hazard Assessment Handbook (CG-446-3); Vol. 4, Response Methods Handbook (CG-446-4).

U.S. Department of Transportation (DOT). 1990. Emergency Response Guidebook. DOT P5600.5, U.S. DOT, Office of Hazardous Materials Transportation, Washington, DC. [Information on potential hazards of DOT regulated hazardous chemicals; updated every three years]

U.S. Environmental Protection Agency (EPA). 1985. Chemical, Physical and Biological Properties of Compounds Present at Hazardous Waste Sites. EPA/530/SW-89-010 (NTIS PB88-224829).

U.S. Environmental Protection Agency (EPA). 1986. Test Methods for Evaluating Solid Waste, 3rd ed. EPA/530/SW-846 (NTIS PB88-239223); First update, 3rd ed. EPA/530/SW-846.3-1 (NTIS PB89-148076). [Current edition and updates available on a subscription basis from U.S. Government Printing Office, Stock #955-001-00000-1]

U.S. Environmental Protection Agency (EPA). 1987. Compendium of Superfund Field Operating Methods. EPA-540/P-87-001, OSWER Directive 9355.0-14 (NTIS PB88-181557), 644 pp.

U.S. Environmental Protection Agency (EPA). 1988. Superfund Exposure Assessment Manual. EPA/540/1-88/001, OSWER Directive 9285.5-1 (NTIS PB90-135859).

U.S. Environmental Protection Agency (EPA). 1989a. Guide for Conducting Treatability Studies Under CERCLA. EPA 540/2-89/058.

U.S. Environmental Protection Agency (EPA). 1989b. Determining Soil Response Action Levels Based on Potential Contaminant Migration to Ground Water: A Compendium of Examples. EPA/540/2089/057.

U.S. Environmental Protection Agency (EPA). 1990. Assessing the Geochemical Fate of Deep-Well-Injected Hazardous Waste: A Reference Guide. EPA 625/6-89-025a. Available from CERI*. [Appendix B provides an index of over 90 references that provide data on sorption and/or biodegradation of more than 150 organic compounds]

U.S. Environmental Protection Agency (EPA). 1992. Handbook of RCRA Ground-Water Monitoring Constituents: Chemical and Physical Properties (Appendix IX to 40 CFR part 264). EPA/530-R-92-022, 267 pp. Office of Solid Waste, Washington, DC.

U.S. Geological Survey. 1982. National Handbook of Recommended Methods for Water Data Acquisition, Chapter 8 (Evaporation and Transpiration). USGS Office of Water Data Coordination, Reston, VA.

van Genuchten, M.Th. and P.J. Wierenga. 1986. Solute Dispersion Coefficients and Retardation Factors. In: Methods of Soil Analysis, Part 1, 2nd ed., A. Klute (ed.), Agronomy Monograph No. 9, American Society of Agronomy, Madison, WI, pp. 1025-1054.

Verschueren, K. 1983. Handbook of Environmental Data on Organic Chemicals, 2nd ed. Van Nostrand Rennhold, New York, 1310 pp. [Data on more than 1,300 organic chemicals.]

Vogel, W.G. 1987. A Manual for Training Reclamation Inspectors in the Fundamentals of Soils and Revegetation. Soil and Water Conservation Society, Ankeny, IA, 178 pp.

Wagenet, R.J. 1986. Water and Solute Flux. In: Methods of Soil Analysis, Part 1, 2nd ed., A. Klute (ed.), Agronomy Monograph No. 9, American Society of Agronomy, Madison, WI, pp. 1055-1088.

Walker, M.M. and L.W. Keith. 1992. EPA's Pesticide Fact Sheet Database. Lewis Publishers, Chelsea, MI. [Manual and 2-3.5"/4-5.25" diskettes containing comprehensive source of information on several hundred pesticides and formulations]

Whittig, L.D. and W.R. Allardice. 1986. X-Ray Diffraction Techniques. In: Methods of Soil Analysis, Part 1, 2nd ed., A. Klute (ed.), Agronomy Monograph No. 9, American Society of Agronomy, Madison, WI, pp. 331-362.

Wilson, L.G. 1982. Monitoring in the Vadose Zone, Part II. Ground Water Monitoring Review 2(4):31-42.

Wischmeier, W.H. and D.D. Smith. 1978. Predicting Rainfall Erosion Losses. Agricultural Handbook 537. U.S. Department of Agriculture, Washington, DC.

Wolbach, C.D., R.R. Whitney, and U.B. Spannegel. 1984. Design and Development of a Hazardous Waste Reactivity Testing Protocol. EPA-600/2-84-057 (NTIS PB84-158807), 143 pp. [Field test kit for on-site compatibility testing of wastes]

Woodruff, N.P. and F.H. Siddoway. 1965. A Wind Erosion Equation. Soil Sci. Soc. Am. Proc. 29:602-608.

Yaws, C., H-C. Yang, and X. Pan. 1991. Henry's Law Constants for 363 Organic Compounds in Water. Chemical Engineering 98(11):179-185.

* ORD Publications, U.S. EPA Center for Environmental Research Information (CERI), P.O. Box 19963, Cincinnati, OH, 45268-0963 (513/569-7562).

APPENDIX D

DESCRIPTION AND INTERPRETATION OF REDOXIMORPHIC SOIL FEATURES

Contaminants will generally enter the ground water most quickly when soils have **aquic** conditions (i.e., experience continuous or periodic saturation and reduction within two meters of the ground surface). Consequently, accurate identification of aquic soil conditions is of special interest when characterizing contaminated sites.

In 1992 the Soil Conservation Service of the U.S. Department of Agriculture adopted a greatly revised, and improved, approach to description and interpretation of aquic soil conditions (Soil Survey Staff, 1992). Previously, mottles and low chroma colors, indicative of reduced iron, (Section 3.1.3) were used as the primary indicator of soil wetness. Better understanding of oxidation-reduction processes has led to more precise methods for describing what are now called **redoximorphic** soil features (i.e. morphologic features indicative of oxidation-reduction processes in the soil).

Three types of saturation are also now recognized in the classification of wet soils: (1) **endosaturation** (all layers from the upper boundary of saturation to a depth of 2 meters are saturated), (2) **episaturation** (one or more unsaturated layers exist below the upper boundary of saturation to a depth of 2 meters—i.e. a perched water table); and (3) **anthric saturation** (a variant of episaturation associated with controlled flooding for such crops as rice and cranberries).

Three separate properties should be identified to define aquic conditions for a soil series: (1) depth and seasonal duration of saturation, (2) occurrence of reduction (Section 3.3.5 and ferrous iron test in Section 3.3.8), and (3) presence of redoximorphic features. When saturation, reduction and redoximorphic features are confirmed for a soil at one site, the saturation and reduction data can be extrapolated to other similar soils using the redoximorphic features alone.

Describing Redoximorphic Features

Redoximorphic features are broadly classified as (1) **redox concentrations** (zones of apparent accumulation of Fe-Mn oxides), (2) **redox depletions** (zones of low chroma—2 or less—where Fe-Mn oxides or both Fe-Mn oxides and clay have been stripped out); and (3) **reduced matrices** with a uniformly low chroma (2 or less) because of the presence of reduced iron.

Redox concentrations come in three main forms: (1) **nodules** and **concentrations** (firm irregularly shaped bodies with diffuse boundaries if formed in situ, or with sharp boundaries after pedoturbation), (2) **masses** (soft bodies of variable shapes within the matrix--features that formerly would have been called "reddish mottles"), and (3) **pore/channel linings** (zones of accumulation that may be either coatings on a pore surface or impregnations of the matrix adjacent to the pore or channel.

Redox depletions consist of features of low chroma (≤ 2) having values of 4 or more and have two major forms: (1) **iron depletions** (low chroma bodies with clay contents similar to that of the adjacent matrix--features that formerly have been called "gray mottles", "gley mottles", "albans" or "neoalbans", and (2) **clay depletions** (low chroma bodies with reduced Fe/Mn and less clay than the adjacent matrix—features that formerly have been described as "silt coatings" or "skeletans".

Form 3-1 contains space for description of redoximorphic soil features. The following abbreviations can be used to describe redox concentrations:

Composition/Type

Fe — iron (see Section 3.3.8 for identification)
Mn — manganese (see Section 3.3.8 for identification)
n — nodules/concretions (concentric rings absent/present)
m — masses
pl — pore linings
lc — linear coatings (vertical or horizontal)

Color (see Section 3.1.3)

Abundance (percent or abbreviations below)

f — few (<2%)
c — common (2-20%)
m — many (>20%)

Size

1 — fine (<5 mm)
2 — medium (5-15 mm)
3 — coarse (>15 mm)

Contrast (see Figure 3-4)

f — faint (1 or 2 units H,C,V)
d — distinct (2-4 units H,C,V)
p — prominent (4-5 units H,C,V)

Location (inped/matrix, exped—along macropores)

Boundaries (sharp—like a knife edge; clear—color grade over <2 mm; diffuse—color grade over >2 mm)

Redox depletions are described in a similar manner as to type (Fe/Mn depletions or clay depletions), color, abundance, size, contrast, and whether the depletions occur along macropores in within the matrix. Redox depletions should be differentiated from illuvial sand and silt coatings and infillings. Sand and silt coatings/fillings typically have a color identical to the of an overlying A or E horizon.

A reduced matrix has uniformly low chroma in situ, but undergoes a change in hue or chroma within 30 minutes after that soil material has been exposed to air. In the absence of visible redoximorphic features, a positive reaction to an $\alpha \alpha'$-dipyridyl solution (Section 3.3.8) qualifies as a redoximorphic feature.

Interpretation of Redoximorphic Features

Inference of magnitude of fluctuations and length of seasonal saturation from redoximorphic features requires corre-lation of features with field measurements of saturation. Nevertheless, a number of general patterns of distribution of redoximorphic features provide insight into how water moves through the soil (Vepraskas, 1992; Fanning et al., 1992):

1. **Redox depletions around macropores and redox concentrations within matrix.** In this situation brown colors are dominant in the matrix and gray colors are dominant on ped or macropore faces. This arrangement of features indicates that water infiltrates the horizon along macropores and reducing condition develop primarily within the macropores. Air is trapped within the matrix and saturation is not long enough to water to enter the matrix or for water to reside within the matrix long enough for reduction to occur. This morphology occurs in soils with perched saturated layers, and is found within and above the slowly permeable layer. It also occurs in fully saturated soils where macropores extend to depth of 2 m or more.

2. **Redox concentration around macropores and redox depletions within matrix.** In this situation gray colors are dominant in the matrix and bright colors dominant on ped or macropore faces. This arrangement of features indicates that the matrix is saturated for periods long enough for reducing conditions to be maintained. Macropores become aerated while the matrix is still saturated so movement of water and reduced substances is from the matrix toward the macropores. Three types of conditions favor this type of morphology: (1) the capillary fringe above a permanent water table, (2) extended perching of a water table above a permeable layer of coarser texture followed by drainage of macropores, and (3) a saturated horizon where plant roots transport air to soil around the roots. Table D-1 provides estimates of the height of the capillary fringe above the zone of saturation based on soil texture.

Table D-1 Guide for Estimation of Capillary Fringe (Mausbach, 1992, based on personal communication, Otto Baumer, 1990)

USDA Texture Class*	Est. Capillary Fringe (cm)
Coarse sand	1-7
Sand	1-9
Fine sand	3-10
Very fine sand	4-12
Loamy coarse sand	5-14
Loamy sand	6-14
Loamy fine sand	8-18
Coarse sandy loam	8-18
Loamy very fine sand	10-20
Sandy loam	10-20
Fine sandy loam	14-24
Very fine sandy loam	16-26
Loam	20-30
Silt loam	25-40
Silt	35-50
Sandy clay loam	20-30
Clay loam	25-35
Silty clay loam	35-55
Sandy clay	20-30
Silty clay	40-60
Clay	25-40

* See Figures 3-1 and 3-2 for texture definitions

3. **Redoximorphic features have no consistent relationship to macropores.** No consistent relationship between redoximorphic features and macropores indicates that reduction and oxidation occur randomly through the horizon over time. This can occur in sands or materials with small aggregates where macropores are either not stable or are relatively small and closely spaced, resulting in air and water using different pathways to move through both the matrix and macropores after each infiltration event.

4. **Redoximorphic features form at the contact between fine- and coarse-textured layers.** In recent floodplain sediments where coarse- and fine-textured materials are stratified, redoximorphic concentrations may accumulate in the coarser strata at the boundaries with finer strata (Fanning et al., 1992). The mechanism for these linear features appears to be iron depletion of the fine-textured material, which tends to have higher concentrations of both free iron and organic matter (for an energy source). The dissolved iron diffuses to the coarser strata where the iron is oxidized by air-filled pores, or more oxygenated ground water.

5. **Fe/Mn Nodules and concretions present.** Iron- and manganese-oxide enriched nodules and concretions, when present, tend to be distributed uniformly through a horizon. They tend to form in heavily leached soils, such as Ultisols, in the upper portion of a seasonally perched water table (Fanning et al., 1992).

Interpretation Problems

Certain soil conditions inhibit the development of redoximorphic features. Evidence of any of the following conditions requires special care when evaluating the presence or absence of aquic conditions:

1. **Low dissolved organic carbon.** Manganese and iron reduction are microbially mediated and require organic carbon for an energy source. In general, dissolved organic carbon concentrations < 10 mg/L will inhibit development of redoximorphic features in saturated soil (Daniels and Buol,

1992). For example, this appears to be the reason that heavily leached tropical soils, which may be saturated for three months or longer, do not exhibit redoximorphic features (Couto et al., 1985). Conversely, unusually high concentrations of dissolved organic carbon could result in more rapid reduction of iron for the same period of saturation compared to saturated soils with less dissolved organic carbon.

2. **Cold soil temperature.** Soil temperature must be warm enough to allow microbial growth. Iron reduction will not occur when soil and ground-water temperature is less than < 5 °C). In temperate climates winter subsurface temperatures may go low enough to inhibit reduction during winter, and very cold climates (frigid, cryic, and pergelic soil temperature regimes—see Soil Survey Staff, 1992, for definitions) year-round soil temperatures may allow minimal to no reduction.

3. **High dissolved oxygen.** Ground-water must be depleted of oxygen for reduction to occur. Reduction will not occur if ground-water in the soil is sufficiently oxygenated to support aerobic microbial activity in the presence of dissolved organic carbon. A soil that is periodically saturated but not reduced is called **oxyaquic.**

4. **Low iron concentrations.** Redoximorphic features will not develop in saturated soils that very low concentrations of iron. This may occur in certain coarse-textured wet soils (Fanning et al., 1992).

5. **High pH.** Critical redox potentials needed for Fe and Mn reduction are pH-dependent, and will not occur at pHs greater than 8. Figure D-1 shows the Eh-pH conditions required for reduction of iron. Collins and Buol (1970) provide a similar Eh-pH diagram for manganese.

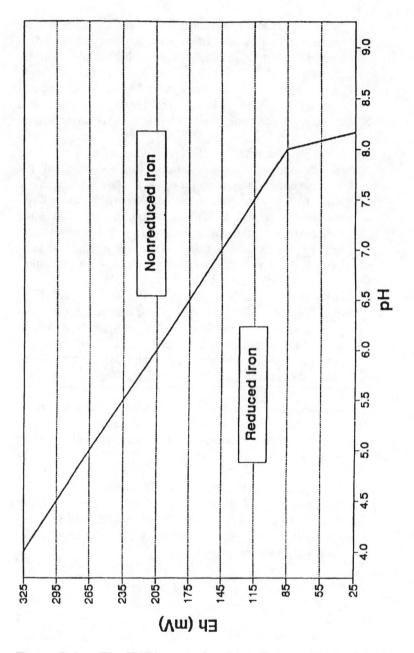

Figure D-1. Eh-pH Diagram for Iron (Source: Hudnall and Wilding, 1992).

When a previously well-drained soil becomes saturated for extended periods, redoximorphic features can develop relatively quickly. For example, Vepraskas and Guertal (1992) estimated that a 0.2 cm depletion coating can form around a 0.1 cm root channel in a matter of years with dissolved Fe(II) concentrations of 10 ppm.

On the other hand, **relict** (or fossil) redoximorphic features may persist for many years after natural or artificial conditions reduce or eliminate saturated conditions in an aquic soil. Criteria for identification of relict redoximorphic features are not well established (Vepraskas, 1992). When Fe-Mn nodules and concretions are actively forming, boundaries with the matrix will tend to be gradual or diffuse. Sharp boundaries, on the other hand, tend to result when nodules or concretions are degrading. Evidence of relatively recent geomorphic events, such as rapid downcutting of stream channels (or artificial drainage systems) may provide clues to changes in soil moisture regime that allow classification of redoximorphic features as relict.

References

Collins, J.F. and S.W. Buol. 1970. Effects of Fluctuations in the Eh-pH Environment on Iron and/or Manganese Equilibria. Soil Science 110:111-118.

Couto, W., C. Sanzonowicz, and A. de O. Barcellos. 1985. Factors Affecting Oxidation-Reduction Processes in an Oxisol with a Seasonal Water Table. Soil Sci. Soc. Am. J. 49:1245-1248.

Daniels, R.B. and S.W. Buol. 1992. Water Table Dynamics and Significance to Soil Genesis. In: Proc. 8th Int. Soil Correlation Meeting (VIII ISCOM): Characterization, Classification and Utilization of Wet Soils, J.M. Kimble (ed.), USDA, Soil Conservation Service, National Soil Survey Center, Lincoln, NE, pp. 66-74.

Fanning, D.S., M.C. Rabenhorst, and M.L. Thompson. 1992. Micro-Macromorphology of West Soils in Relation to Classification. In: Proc. 8th Int. Soil Correlation Meeting (VIII ISCOM): Characterization, Classification and Utilization of Wet Soils, J.M. Kimble (ed.), USDA, Soil Conservation Service, National Soil Survey Center, Lincoln, NE, pp. 106-112.

Hudnall, W.H. and L.P. Wilding. 1992. Monitoring Soil Wetness Conditions in Louisiana and Texas. In: Proc. 8th Int. Soil Correlation Meeting (VIII ISCOM): Characterization, Classification and Utilization of Wet Soils, J.M. Kimble (ed.), USDA, Soil Conservation Service, National Soil Survey Center, Lincoln, NE, pp. 135-147.

Mausbach, M.J. 1992. Soil Survey Interpretations for Wet Soils. In: Proc. 8th Int. Soil Correlation Meeting (VIII ISCOM): Characterization, Classification and Utilization of Wet Soils, J.M. Kimble (ed.), USDA, Soil Conservation Service, National Soil Survey Center, Lincoln, NE, pp. 172-178.

Vepraskas, M.J. 1992. Redoximorphic Features for Identifying Aquic Conditions. North Carolina Agricultural Research Service Technical Bulletin 301, 33 pp. Department of Agricultural Communications, Box 7603, North Carolina State University, Raleigh, NC 27695-7603 ($5.00).

Vepraskas, M.J. and W.R. Guertal. 1992. Morphological Indicators of Soil Wetness. In: Proc. 8th Int. Soil Correlation Meeting (VIII ISCOM): Characterization, Classification and Utilization of Wet Soils, J.M. Kimble (ed.), USDA, Soil Conservation Service, National Soil Survey Center, Lincoln, NE, pp. 307-312.

Soil Survey Staff. 1992. Keys to Soil Taxonomy, 5th ed. SMSS Technical Monograph No. 19. Pocahontas Press, P.O. Drawer F, Blacksburg, VA, 24063-1020, 541 pp. ($20.00 plus $2.50 postage and handling; 5% discount for prepaid orders).

APPENDIX E

COMPILATION OF ASTM STANDARDS RELATED TO SAMPLING AND MEASUREMENT OF SOIL PHYSICAL AND CHEMICAL PROPERTIES

This Appendix lists more than 90 standard test methods, practices, and guides developed by the American Society for Testing and Materials (ASTM) for description, sampling and measurement of soil physical and chemical properties covered in this Guide. Appendix C (Tables C-1 through C-5) identify methods for measuring specific physical or chemical properties by ASTM's alphanumeric designation.

This appendix simply lists ASTM's alphanumeric designation and the full title of the test method, practice or guide. A **standard test method** is a definitive procedure for the identification, measurement and evaluation of one or more qualities, characteristics, or properties of a material or system that produces as a test result. A **standard practice** is a definitive procedure for performing one or more specific operations or functions that does not produce a test result. A **standard guide** offers a series of options or instructions, but does not recommend a specific course of action.

ASTM alphanumeric designations with dash indicate the year that the standard was established, or the year of the most recent significant revision to the standard. Not all standards in this appendix include this information because not all sources used in compiling the appendix provided this information. Knowing the year a standard was established or revised is helpful, because it indicates whether an earlier editions of the annual ASTM book of standards containing a particular standard can be used. For example, any edition of Volume 4.08 of ASTM's Annual Book of Standard that has been published since 1963 can be used for D422-63 (Test Method for Particle-Size Analysis of Soils). However, if editions published prior to the current year are used, ASTM should be contacted to make sure that a particular standard of interest has not been revised.

Most ASTM standards related to soil are published in Volume 4.08 (Soil and Rock; Dimension Stone; Geosynthetics). This volume, and individual standards, can be purchased from: ASTM, 1916 Race Street, Philadelphia, PA 19103-1187 (215/299-5400).

E-1

ASTM Standards (In Volume 4.08, unless otherwise specified)

C998-83   Method for Sampling Surface Soil for Radionuclides (Vol. 11.04).

D421-85   Practice for Dry Preparation of Soil Samples for Particle-Size Analysis and Determination of Soil Constants

D422-63   Method for Particle-Size Analysis of Soils.

D427-83   Test Method for Shrinkage Factors of Soils.

D698-91   Test Method for Laboratory Compaction Characteristics of Soil Using Standard Effort.

D854-83   Test Method for Specific Gravity of Soil.

D932   Test Method For Iron Bacteria in Water and Water-Formed Deposits (Vol. 11.02).

D1140-92   Test Method for Amount of Materials in Soils Finer than No. 200 (75-μ) Sieve.

D1194-72   Test Method for Bearing Capacity for Soil for Static Load and Spread Footings.

D1452-80   Practice for Soil Investigation and Sampling by Auger Borings.

D1498-76   Practice of Oxidation-Reduction Potential of Water (Vol. 11.02).

D1556-90   Test Method for Density of Soil In Place by the Sand-Cone Method.

D1557-91   Test Method for Laboratory Compaction Characteristics of Soil Using Modified Effort.

D1586-84   Method for Penetration Test and Split-Barrel Sampling of Soils.

D1587-83   Practice for Thin-Walled Tube Sampling of Soils.

E-2

D1883-87 Test Method for CBR (California Bearing Ratio) of Laboratory-Compacted Soils.

D2166-91 Test Method for Unconfined Compressive Strength of Cohesive Soil.

D2167-84 Test Method for Density and Unit Weight of Soil In Place by the Rubber Balloon Method.

D2216-90 Standard Test Method for Laboratory Determination of Water (Moisture) Content of Soil and Rock. [Gravimetric oven drying]

D2217-85 Practice for Wet Preparation of Soil Samples for Particle-Size Analysis and Determination Soil Constants.

D2325-68 Test Method for Capillary-Moisture Relationship for Coarse- and Medium-Textured Soils by Porous-Plate Apparatus. [Soil water retention]

D2434-68 Test Method for Permeability of Granular Soils (Constant Head).

D2487-92 Test Method for Classification of Soils for Engineering Purposes.

D2488-90 Practice for Description and Identification of Soils (Visual-Manual Procedures).

D2573-72 Test Method for Field Vane Shear Test in Cohesive Soil.

D2850-87 Test Method for Unconsolidated, Undrained Compressive Strength of Cohesive Soils in Triaxial Compression.

D2922-81 Test Methods for Density of Soil and Soil-Aggregate In Place by Nuclear Methods (Shallow Depth). [Gamma-gamma, surface or < 12")

D2937-83 Test Method for Density of Soil in Place by the Drive-Cylinder Method.

D2974-87 Test Methods for Moisture, Ash, and Organic Matter of Peat and Other Organic Soils.

D2976-71   Test Method for pH of Peat Materials.

D3017-88   Test Method for Water Content of Soil and Rock In Place by Nuclear Methods (Shallow Depth). [Neutron Probe]

D3080-90   Test Method for Direct Shear Test of Soils Under Consolidated Drained Conditions.

D3152-72   Test Method for Capillary-Moisture Relationships for Fine-Textured Soils by Pressure Membrane Apparatus.

D3385-88   Standard Test Method for Infiltration Rate of Soils in Field Using Double-Ring Infiltrometers.

D3404-91   Standard Guide to Measuring Matric Potential in the Vadose Zone Using Tensiometers.

D3441-86   Method for Deep-Quasi-Static, Cone and Friction-Cone Penetration Tests of Soil

D3550-84   Practice for Ring-Lined Barrel Sampling of Soils.

D4012   Test Method for Adenosine Triphosphate (ATP) Content of Microorganisms in Water (Vol. 11.02).

D4030-83   Method of Measuring Humidity with Cooled-Surface Condensation (Dew Point) Hygrometer (Vol. 11.03).

D4053-89   Practice for Description of Frozen Soils (Visual-Manual Procedure).

D4220-89   Practices for Preserving and Transporting Soil Samples.

D4221-90   Test Method for Dispersive Characteristics of Clay Soil by Double Hydrometer.

D4318-84   Test Method for Liquid Limit, Plastic Limit, and Plasticity Index of Soils.

D4319-83   Test Method for Distribution Ratios by the Short-Term Batch Method.

D4373-84   Test Method for Calcium Carbonate Content.

D4404-84   Test Method for Determination of Pore Volumes and Pore Volume Distribution of Soil and Rock by Mercury Intrusion Porosimetry.

D4412   Test Methods for Sulfate-Reducing Bacteria in Water and Water-Formed Deposits (Vol. 11.02).

D4429-84   Test Method for Bearing Ratio of Soils in Place.

D4452-85   Methods for X-Ray Radiography of Soil Samples.

D4454   Test Method for Simultaneous Enumeration of Total Respiring Bacteria in Aquatic Systems by Microscopy (Vol. 11.02).

D4455   Test Method for Enumeration of Aquatic Bacteria by Epifluorescence Microscopy Counting Procedure (Vol. 11.02).

D4478   Test Methods for Oxygen Uptake (Vol. 11.02). [Microbial respiration]

D4480-85   Test Method for Measuring Surface Wind by Means of Wind Vanes and Rotating Anemometers (Vol. 11.03).

D4525-90   Test Method for Permeability of Rocks by Flowing Air.

D4531-86   Test Method for Bulk Density of Peat and Peat Products.

D4542-85   Test Method for Pore Water Extraction and Determination of the Soluble Salt Content of Soils by Refractometer.

D4546-90   Test Methods for One-Dimensional Swell or Settlement Potential of Cohesive Soils.

D4547-91   Practice for Sampling Waste and Soils for Volatile Organics (Vol. 11.04)

D4564-86   Test Method for Density of Soil In Place by the Sleeve Method. [Cohesionless, gravelly soils]

D4633-86   Test Method for Stress Wave Energy Measurement for Dynamic Penetrometer Testing Systems.

D4643-87   Test Method for Determination of Water (Moisture) Content) of Soil by the Microwave Oven Method.

D4646-87   Test Method for 24-Hour Batch Type Measurement of Contaminant Sorption by Soils and Sediments (Vol 11.02).

D4647-87   Test Method for Identification and Classification of Dispersive Clay Soils by the Pinhole Test.

D4648-87   Test Method for Laboratory Miniature Vane Shear Test for Saturated Fine-Grained Clayey Soil.

D4696-92   Guide for Pore-Liquid Sampling From the Vadose Zone.

D4700-91   Guide for Soil Sampling from the Vadose Zone.

D4750-87   Test Method for Determining Subsurface Liquid Levels in a Borehole or Monitoring Well (Observation Well).

D4767-88   Test Method for Consolidated-Undrained Triaxial Compression Test on Cohesive Soils.

D4914-89   Test Methods for Density of Soil and Rock In Place by the Sand Replacement Method in a Test Pit. [Soils with particles larger than 3 inches]

D4943-89   Test Method for Shrinkage Factors of Soils by the Wax Method.

D4944-89   Test Method for Field Determination of Water (Moisture) Content of Soil by the Calcium Carbide Gas Pressure Tester Method.

D4959-89   Test Method for Determination of Water (Moisture) Content of Soil by Direct Heating Method.

D4972-89   Test Method for pH of Soils.

D4982-89   Method for Flammability Potential Screening Analysis of

Waste (Vol. 11.04).

D5057-90   Method for Screening of Apparent Specific Gravity and
            Bulk Density of Waste (Vol. 11.04).

D5079-90   Practices for Preserving and Transporting Rock Core
            Samples.

D5080-90   Test Method for Rapid Determination of Percent
            Compaction.

D5088-90   Practice for Decontamination of Field Equipment Used at
            Nonradioactive Waste Sites.

D5093-90   Standard Test Method for Field Measurement of
            Infiltration Rate Using a Double-Ring Infiltrometer With
            a Sealed-Inner Ring

D5126-90   Guide for Comparison of Field Methods for Determining
            Hydraulic Conductivity in the Vadose Zone.

D5195-91   Test Method for Determination of Density of Soil and
            Rock In-Place at Depths Below the Surface by Nuclear
            Methods. [Gamma-gamma, > 12"]

D5220-92   Test Method for Water Content of Soil and Rock In-Place
            by the Neutron Depth Probe Method.

D5298-92   Test Method for Measurement of Soil Potential (Suction)
            Using Filter Paper.

E337-84   Test Method of Measuring Humidity with a Psychrometer
            (the Measurement of Wet- and Dry-Bulb Temperatures)
            (Vol. 11.03).

E816-84   Method for Calibration of Secondary Reference
            Pyrheliometers and Pyrheliometers for Field Use (Vol.
            11.03).

E824   Method for Transfer of Calibration from Reference to
            Field Pyranometers (Vol. 11.03).

E895   Practice for Determination of Hydrolysis Rate Constants of Organic Chemicals in Aqueous Solutions (Vol. 11.02).

E1109   Test Method for Determining the Bulk Density of Solid Waste Fractions (Vol. 11.04).

E1147   Test Method for Partition Coefficient (n-Octanol/Water) Estimation by Liquid Chromatography (Vol. 11.02).

E1148   Test Method for Measurements of Aqueous Solubility (Vol. 11.04).

E1194   Test Method for Vapor Pressure (Vol. 11.04).

E1195-87   Test for Sorption Constant (Koc) for Organic Chemicals in Soil and Sediments (Vol. 11.02).

E1196   Test Method for Determining the Anaerobic Biodegradation Potential of Organic Chemicals (Vol. 11.02).

E1279   Test Method for Biodegradation by a Shake-Flask Die-Away Method (Vol. 11.02).

G51-77   Test Method for pH of Soil for Use in Corrosion Testing.

G57-78   Method for Field Measurement of Soil Resistivity Using the Wenner Four-Electrode Method.